Integrated Periodization in Sports Training & Athletic Development

Tudor Bompa / Boris Blumenstein (Ed.) / James Hoffmann / Scott Howell /
Iris Orbach (Ed.)

INTEGRATED PERIODIZATION IN SPORTS TRAINING & ATHLETIC DEVELOPMENT

Combining Training Methodology, Sports Psychology, and Nutrition to Optimize Performance

Meyer & Meyer Sport

British Library Cataloguing in Publication Data
A catalogue record for this book is available from the British Library

Integrated Periodization in Sports Training & Athletic Development
Maidenhead: Meyer & Meyer Sports (UK) Ltd., 2019
ISBN: 978-1-78255-141-6

Aachen, Auckland, Beirut, Cairo, Cape Town, Dubai, Hägendorf, Hong Kong, Indianapolis, Manila, New Delhi, Singapore, Sydney, Tehran, Vienna

Member of the World Sports Publishers' Association (WSPA), www.w-s-p-a.org

CREDITS

Cover design: Anja Elsen
Interior design: Annika Naas
Layout: Amnet
Cover images: © AdobeStock
Interior figures: © Tudor Bompa, Boris Blumenstein, James Hoffmann, Scott Howell, and Iris Orbach, unless otherwise noted
Managing editor: Elizabeth Evans
Copyediting: Amnet

Printed by C-M Books, Ann Arbor, MI, USA
ISBN: 978-1-78255-141-6
Email: info@m-m-sports.com
www.thesportspublisher.com

CONTENTS

PREFACE

Athletes training and their participation in competitions must be seen as intricate elements of human behavior.

During training the body and mind are modeled to behave in a specific way during competition. While the body is modeled by specific training methods, the mind is also perfected by specific techniques, using the energy supplied by best nutritional plans.

Integrated periodization (IP) illustrates how to integrate in training the main aspects of physical, psychological, and nutrition plans. However, physical training is succinctly presented in this book since two of Tudor Bompa's other books refer in detail about it. Therefore, physical training in this book must be viewed just as a guiding element on which basis nutritional and psychological programs are planned for specific sport disciplines.

Chapter 1

PRESENT STATE OF THE ART

Tudor Bompa, PhD / Boris Blumenstein, PhD / Iris Orbach, PhD / James Hoffman, PhD

INTRODUCTION

For far too long, research in sports science and the methodology of training have focused, often in isolation, on their specific areas of concern. Researches in exercise physiology, sports psychology, and the science of nutrition have evolved separately, without considering that their focus should be on those who could benefit from them: the athlete and coach! And the athlete is a complex being with specific requirements and in need of professional/ scientific services.

To be effective and meet the needs of the athletes, sports sciences must consider the specific objectives of training and the manner in which training is planned and periodized. However, some of these sports scientists are seldom aware that athletes and coaches need their help throughout the year! Not just before a competition. Or after failing to meet planned performance/testing objectives when some scientists may be needed to heal eventual psychological wounds.

Furthermore, an athlete involved in competitions may reach the objectives much easier when sports scientists and coaches collaborate together. Under these conditions human physiological potentials can be drastically improved and overcome athletes' limitations. However, athletic improvements cannot be realized without a complex fueling of the body and mind.

Recognizing the complex needs of the competing athletes, the first proposal to integrate all aspects of sport training was made and registered as an intelligent property in 1997 and published in 1999 (Bompa, 1999). An integrated plan has been suggested along with periodization of motor abilities, nutrition plans to use, and the kind of psychological strategies to develop and refine. However, latest research, especially in nutrition, has changed quite visibly. As a result, our suggested models regarding *integrated periodization* made for specific sports do incorporate everything what is new in the science of nutrition and sports psychology.

PERIODIZATION AND INTEGRATED PERIODIZATION DEFINED

Already accepted as a classical term in the science and methodology of training, *periodization* is the foundation of planning the annual training programs for athletes. The etymology of the term originates from the word *period*, which refers to a specific division of time. In the theory of training, it refers to a *phase* of training. This is why periodization is often referred to as phase-based training. Finally, in sports training, periodization is regarded as a method by which training is divided into smaller phases, easier to manage training programs.

The term *periodization* has been, however, borrowed from other sciences, such as history (antiquity, medieval history, etc.) or literature (Shakespearian, Victorian literature, etc.).

As a training concept, periodization, not exactly in the present form, has existed for long period of times. Although its origin is unknown, in its unrefined form it, has existed since the ancient Olympic Games (776–393 BC). Several manuals on planning and training have been written by Flavius Philostratus (AD 170–245). According to Philostratus, Greek Olympians have used an annual plan with the following periods:

1. *Preparatory* phase for the Olympic Games, where some informal competitions have been planned
2. *Competitive* phase, or the Olympic Games
3. *Rest* period

Conceptually, what most coaches do nowadays is not much different than the ancient Olympians except that our sophistication is based on advancements in science and methodology and that the competitive phase in many sports is longer and with higher number of competitions.

As a complex training concept, periodization does not refer only to how an annual plan is divided into different phases (figure 1.1), but most importantly, it refers to the periodization of motor abilities (figure 1.2) such as speed, strength, power, agility, and endurance.

<table>
<tr><td></td><td colspan="27">Annual Plan</td></tr>
<tr><td>Training phase</td><td colspan="7">Preparatory</td><td colspan="16">Competitive</td><td colspan="4">Transition</td></tr>
<tr><td>Macro cycles</td><td colspan="3"></td><td colspan="4"></td><td colspan="3"></td><td colspan="3"></td><td colspan="3"></td><td colspan="3"></td><td colspan="4"></td><td colspan="4"></td></tr>
<tr><td>Micro cycles</td><td></td><td></td><td></td><td></td><td></td><td></td><td></td><td></td><td></td><td></td><td></td><td></td><td></td><td></td><td></td><td></td><td></td><td></td><td></td><td></td><td></td><td></td><td></td><td></td><td></td><td></td><td></td></tr>
</table>

Figure 1.1 Periodization of training phases during an annual plan

Training phase	Preparatory		Competitive		Transition
Periodization of strength	AA	MxS	P/A	Maintenance of power	AA

Legend: AA = Anatomical adaptation MxS = Maximum strength P/A = Power/agility

Figure 1.2 Periodization of strength/power during an annual plan

If we appeal to our imagination, we may consider Philostratus also as the early proponent of *integrated periodization*, since he proposed that a coach must also be a *psychiatrist*!

Another proponent of close relationships between *mind and body* was the Roman physician Claudius Galenus, or Galen (AD 129–217), who in his book *The Art of Preservation of Health* proposed that those who exercise should also have a *good nutrition* and that after exercising one must have a bath to *relax* the body and mind. Equally important, Galen suggested how to treat psychological problems using *talk therapy* so that the individual can share his secrets and passions.

While Galen and Philostratus have recognized the complex qualities needed by athletes and coaches (training, nutrition, and psychiatry), some 1,800 years later, experts in the field are still working in isolation from each other! Although nutrition experts and psychologists work directly with some athletes, most athletic needs are not yet *integrated* year-round.

Not only that but sports scientists consult athletes often in disregard of periodization of training and the objectives of each training phase. Unless coaches, sports scientists, and athletes work together to meet the goals of a specific training phase, athletes' performance may not be maximized. This is the reason we must *integrate* all the aspects that are needed in sports training. Therefore, we have taken the audacity of proposing this book to you, the reader.

CONFUSION BETWEEN PERIODIZATION AND LOADING

Although its existence is over 2,500 years long and its effectiveness constantly demonstrated by top athletes, there are still some individuals who actually question periodization. If periodization is disputed, so should training, planning, and organization or even the physiology of adaptation! Unfortunately, periodization is sometimes confused with loading patterns!

There are some authors and sports scientists who promote a periodization fallacy in the sense that they discuss *linear* and *undulatory* (or *wavelike*) periodization! Sometimes this confusion goes as far as even escaping scientific scrutiny and, as a result, being used in titles of theses for graduate studies! Incredible where an innocent confusion can lead to!

As specified above, periodization strictly refers to phase-based training! And training phases, or periodization, cannot be linear or undulatory! In fact, *periodization* is nothing else but a sequence of *training periods* (therefore the term *periodization*), or phases. Each of these phases has specific training goals with the final scope being to take athletes' potentials to the highest level possible prior to and during the competitive phase.

As a flexible concept of planning, periodization also has several variations from the main model, depending on the specifics of sports and the schedule of competitions an athlete participates per year. There is no linear or undulatory periodization! These two terms refer only to the methodology of loading the training program for an athlete or team. Not to periodization!

While periodization has been briefly defined above, the methodology of loading has several variations.

De Lorme and Watkins (1951) were among the first to refer to progressive loading, which in later years has been transformed into the *linear* method, or the method of constantly and daily increasing the load of training. Linear loading is specifically visible in bodybuilding training, this loading pattern being one of the reasons the athletes in this sport are constantly *overtrained!* The physiological *principle of progressive adaptation* is often disregarded by individuals who promote *linear periodization* and still believe in *more is better*! As for undulatory loading method, this was promoted by Bompa in 1956 in Romania and published in the United States in 1983 (Bompa, 1956).

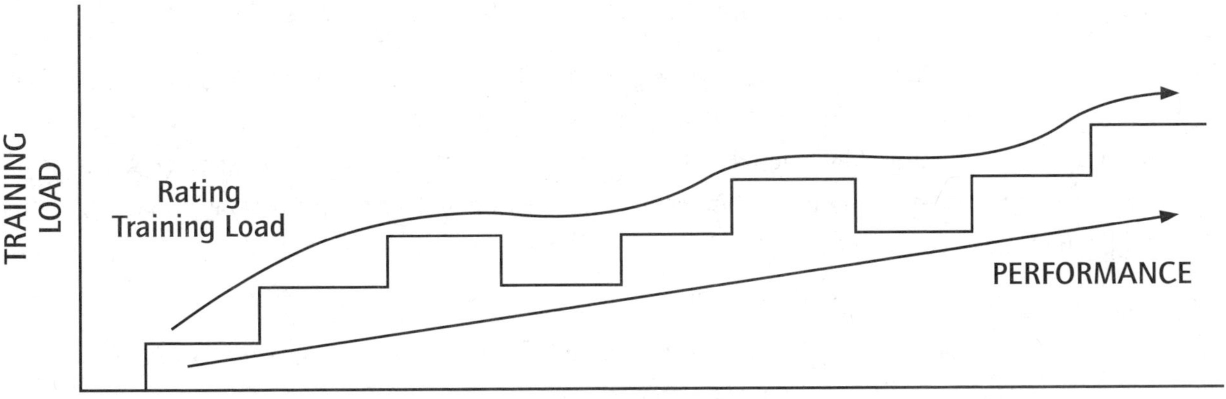

The curve of rating the training load appears to be undulatory while the performance improves continuously (the arrow).

Figure 1.3 Step training load over a time period

Reprinted by permission from T.O. Bompa, Periodization: Theory and Methodology of Training, 4th ed. *(Champaign, IL: Human Kinetics, 1999), 48.*

The progression of the load of training is made in steps. But if you link the top of each step over several cycles, you obtain an undulatory shape, while performance is intended to consistently improve. However, as illustrated by figure 1.3, undulatory loading is strictly the result applying *step loading* over a longer period.

WHY ATHLETE TRAINING MUST BE INTEGRATED WITH NUTRITION AND SPORTS PSYCHOLOGY

Athlete training is often discussed from the point of view of different training methods or as it relates to energy systems but very rarely as a complete and comprehensive concept. Athletes are complex human beings with specific needs. In addition to technical, tactical, and physical training, athletes require very specific nutrition plans and psychological strategies. Yet, professionals in different fields related to athletes training discuss these aspects in isolation from each other, forgetting that at the center of our concerns is the athlete. Therefore, this book discusses the main elements of athletes' and coaches' concerns: training, nutrition, and sports psychology.

NUTRITIONAL PREPARATION IN SPORTS

Within the last 50–60 years or so, many advancements have been made in all things relating to athlete preparedness, including the physical, psychological, and nutrition preparation strategies. Most discussions on periodization have traditionally revolved around the systematic alterations of training variables to maximize performance and concurrently manage stressors throughout a training cycle; however, these discussions are often largely incomplete when they fail to address the impact of sport nutrition and nutritional periodization on these same outcomes. At the same time, the roles of both sport nutrition and nutritional periodization have often been misinterpreted, leading to suboptimal practices. To integrate strategies effectively, a better understanding of these roles is required.

Currently, the traditional model of sports nutrition has been heavily influenced by clinical nutrition practices: lowering cardiovascular disease risk factors, improving micronutrient content, eating colorful fruits and vegetables, and modifying behavior toward a healthier lifestyle. Although these are excellent and valuable strategies, they do not necessarily serve the needs of sport nutrition. Accordingly, coaches and athletes often devalue or underutilize the nutritional resources available to them.

For our purposes sport nutrition is not clinical or health nutrition, and although promoting health is certainly worthwhile, it is not the focus of sport nutrition. Sport nutrition can be boiled down into one simple goal: the enhancement of sport performance through nutritional intervention. We want to win, and anything that is not significantly contributing to this goal can be set aside for later. In terms of integrated periodization, sport nutrition seeks to improve upon the areas of metabolism, body composition alteration, and enhancing recovery-adaptive (RA) processes to boost athlete performance. We must be sure that athletes are being fueled properly to perform well, adjusting their bodyweight and size at appropriate times and rates, and establish good eating habits to allow them to bounce back from hard training and competitions.

ENHANCING METABOLISM

The food we consume literally serves as the fuel for training. The types and quantities of food have a direct impact on a person's ability to perform intense and voluminous exercise. This is true not only on a more generalized scale, such as eating a sufficient number of calories per day to meet exercise demands, but also at the cellular level where the presence or absence of certain macronutrients like carbohydrates can play a huge role in regulating energy metabolism. Therefore, the ability to generate an overloading training stimulus and drive training adaptations, is inherently linked to the capacity to convert the stored potential energy in foodstuffs into the mechanical energy of exercise. Even an annual training plan handed down by Zeus himself will fall short if the athlete is not adequately fueled for training and competition.

Enhancements in metabolism seek to improve not only the total capacity of available energy for training but also the ability to produce energy at a sufficient rate to meet the demands of intense training. This is obviously important for competition scenarios; however, training at a high level of performance is also critical for generating sufficient and progressive overloads over time, to drive the RA processes. Sport nutrition practices seek to optimize performance of competition and exercise through manipulation of total available energy (calorie balance), substrate composition of the energy supply (macronutrients), time that food is eaten relative to exercise, types of foods consumed, and judicious use of sport supplements.

ENHANCING BODY COMPOSITION

All sports have normative standards of performance within the areas of bodyweight, body fat, and muscularity. Although some coaches and scientists disregard this topic as being for "meatheads," bodybuilders, and those only concerned with muscular hypertrophy, the short- and long-term changes made in body composition are an integral and inherent part of training and nutrition periodization. For some sports, this comes in the form of optimizing power to bodyweight ratios, whereas other sports such as those with weight classes benefit immensely from having the lowest possible body fat and highest amount of muscle at the given weight class. Still, other sports simply seek out an individualized optimal competition bodyweight to maximize performance.

This concept of body composition enhancement does include integrative strategies for gaining muscle mass; however, this is a relatively small piece in a large puzzle. Experience has shown that asking athletes to gain or lose weight without addressing what they are actually eating quickly becomes an effort in futility. Training components generate the protein turnover needed to grow/maintain muscle mass as well as provide a productive outlet for energy expenditure, whereas nutritional components ensure that the energy conditions are met and that the body tissues are changing in the most favorable way. For example, when an athlete must lose weight, the goal is to retain as much as muscle mass as possible while reducing fat stores. Conversely, when an athlete must gain weight, the goal is to grow muscle at the highest possible rate while minimizing the accumulation of fat throughout the process. The integration of training and nutrition optimizes the times at which body composition alterations occur throughout the annual plan to maximize these effects and minimize potential detriments. A conceptual outline is represented in figure 1.4.

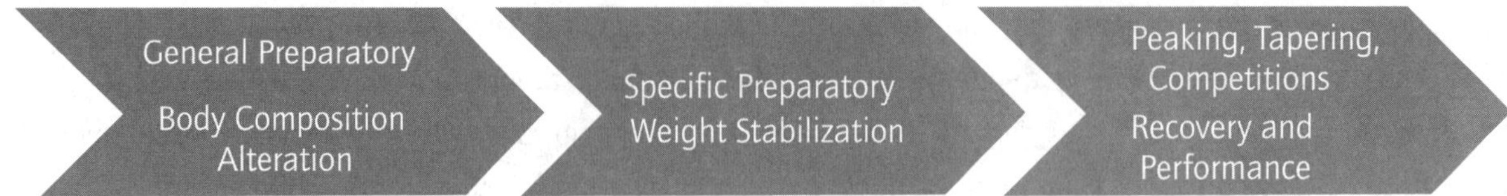

Figure 1.4 Periodized nutrition and training timeline for sport

ENHANCING RECOVERY-ADAPTATION (RA)

Arguably one of the most important components of a sport nutrition program is not only ensuring optimal conditions for training and competing but also ensuring optimal conditions for the subsequent signaling cascades involved in RA. Although training provides the primary stimulus for protein synthesis and altered gene expression, the ability of the athlete to effectively move into recovery periods can play a role in governing the adaptive response. Nutritional strategies play a vital role in promoting recovery from training not only in the immediate postexercise periods but also in the days that follow while protein synthesis remains elevated. Postexercise recovery strategies can positively influence endocrine responses, balancing anabolic and catabolic conditions, cellular signaling pathways, and nervous system activity in addition to the more well-known processes of rehydrating the athlete and replenishing lost carbohydrate stores.

Nutrition provides the literal building blocks to replenish lost glycogen stores and facilitate skeletal muscle protein synthesis, which are both inherently stimulated by exercise itself, making nonintegrated discussions on these topics largely incomplete. Integrated strategies to promote postexercise recovery are essential in adequately preparing athletes for the rigors of training and competition as well as driving the adaptive responses under the most favorable conditions.

This book will guide you on foundational concepts in sport nutrition, the theory behind basic nutritional periodization, and offer practical examples of how to implement these strategies into the real world. Along with the chapters regarding training, allostatic management, and psychology, this book will provide a holistic approach to periodization never explored before and will help coaches, athletes, and scientists bring the total package to all things athletics.

PSYCHOLOGICAL PREPARATION IN SPORTS

Sport psychology has become a major foundation of any sport program that aspires to train elite athletes and prepare them for a competitive event. Training programs for elite athletes utilize several sport-related domains, such as sport medicine, nutrition, physiology, and sport psychology, aiming to achieve maximum success of the athletes as well as the coaching staff. Psychological preparation (PP) for individual athletes as well as team performance is an important factor for athletic achievement; however, it is only one aspect of an athlete's preparation.

The typical structure of sport training is composed of four different types of preparations: physical, technical, tactical, and psychological (Bompa, 1999; Bompa & Haff, 2009). Each of these preparations is interrelated to the other preparations and is based on sport discipline objectives and training principles. The four preparations span across gender, age, and skill level. Moreover, their length, intensity, and nature may vary according to the specific needs of the athlete, the team, or situational circumstances. In this section, we will introduce the readers to the definition and role of PP in sport training: psychological factors associated with athletes' success. In the following chapters, PP will be discussed in a more extensive manner. Research and applied evidence have shown that PP has a profound effect on athletic performance and achievement (e.g., Gould & Maynard, 2009; Vealey, 2007). PP provides athletes with specific psychological techniques/strategies that can help them effectively cope with the mental and emotional barriers that they are likely to encounter during training and competition, leading to peak performance.

PEAK PERFORMANCE

Peak performance in sport is usually associated with personal best results. In modern sports, athletes and coaches realize the importance of the psychological factors that play a role during peak performance at major competitions. Bompa (1999) describes an athlete's peaking as a special training state that is "characterized by a high CNS adaptation, motor and biological harmony, high motivation, ability to cope with frustration, accepting the implicit risk of competing, and high self-confidence" (p. 95). Similarly, sport psychologists have characterized peak performance as a product of the body and mind, which possibly can be trained (Krane & Williams, 2015). A major condition for achieving peak performance in elite sport is for the athletes to learn how to attain and preserve an optimal mental state under competitive stress distractions.

OPTIMAL MENTAL STATE

Sport scientists and practitioners describe the optimal mental state (ideal body/mind state), in which athletes put it all together–both physically and mentally, before and during peak performance. This optimal state has also been referred to as flow state (Csikszentmihalyi, 1990), or individual zone of optimal functioning (Hanin, 2000;Jackson, 2000). An ideal mental state can be accomplished by using the necessary psychological skills that enable athletes to cope with competitive stress and achieve peak performance (figure 1.5).

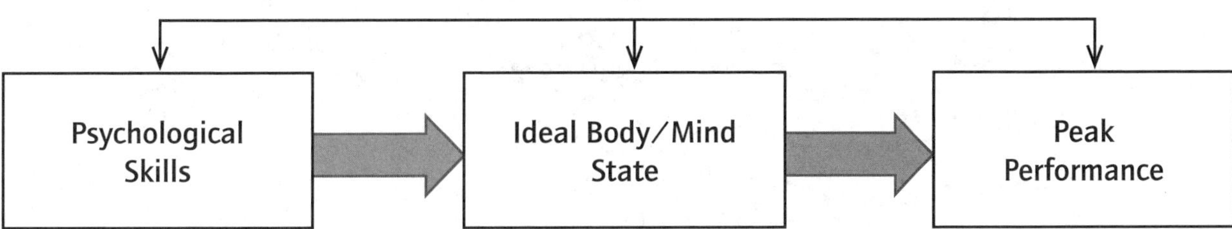

Figure 1.5 Schematic description for achieving peak performance

In elite sport, creating the optimal state or "being in the zone" is a crucial factor that separates winners from losers. In this state, the body functions automatically, with little conscious attention or effort. Additional characteristics of the ideal mental state include high concentration (appropriately focused); self-regulation of arousal (energized yet relaxed, without stressful distractions); high level of self-confidence; being in control, yet relaxed; and determination and commitment to high performance (Krane & Williams, 2015). However, this optimal state does not happen by itself and requires a long and involved training process aimed at learning, developing, and practicing psychological skills and strategies.

PSYCHOLOGICAL SKILLS AND STRATEGIES

Elite athletes have identified their own ideal mental state and have learned to create and maintain this state voluntarily with psychological skills and strategies. There are a number of psychological skills and strategies commonly used by elite athletes in their quest to achieve peak performance: relaxation and arousal regulation, concentration, goal setting, imagery (visualization), self-confidence, and preperformance routines (Blumenstein & Orbach, 2012a,b). Additional psychological skills for team sport are cohesion, leadership, and communication (Vealey, 2007). All psychological skills can be learned and therefore need to be practiced systematically "just like physical skills" (Weinberg & Williams, 2015). These psychological skills and strategies assist athletes "to arrive at an ideal performance state or condition that is related to optimal psychological states and peak performance either for competition or practice" (Gould, Flett, & Bean, 2009, p. 53). Each sport discipline requires a different combination of the dominant psychological skills and strategies, which will be discussed in chapters 3 and 5.

Today, athletes' and coaches' awareness of the potential and role of PP is ambivalent. On one hand, there is agreement that PP is a crucial factor for success. On the other hand, most coaches and athletes mainly practice body improvement and may overlook psychological/mind factors. The last two decades have seen a significant increase in using PP and an integration of PP with the training process (Balague, 2000; Blumenstein, Lidor & Tenenbaum, 2005, 2007; Blumenstein & Orbach, 2012b, 2018; Holliday et al., 2008). Today, most PP is carried out as independent preparation, without taking into consideration the other training factors (figure 1.6).

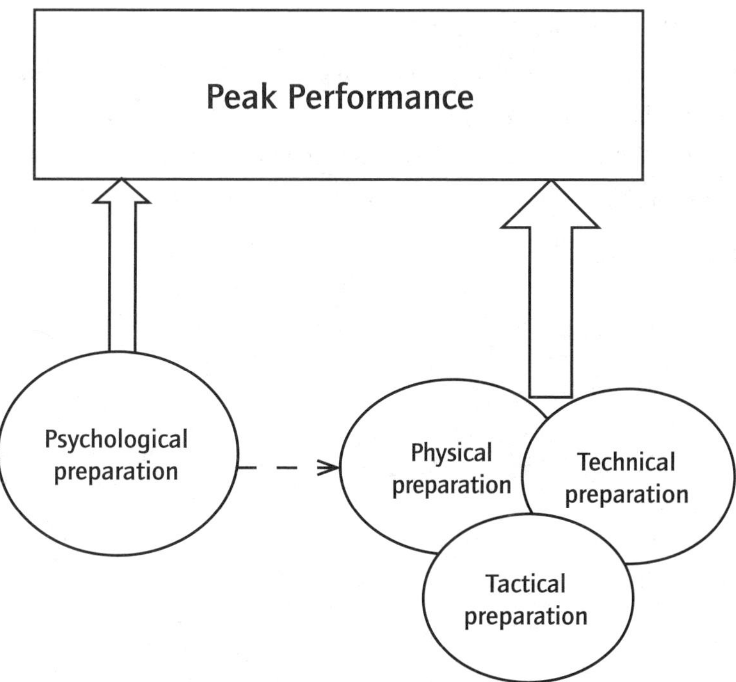

Figure 1.6 Typical position of psychological preparation in the training process

This situation may be due to several reasons:

- There is a misconception that PP is appropriate only for elite sport levels.
- Usually, an athlete refers to a sport psychology consultant before a target competition and in cases of struggling with a conflict/barrier/obstacle, with the expectation that their cooperation will be for a short time period.
- Coaches may not have enough knowledge about PP; therefore they may be skeptical, wait for a quick response, and, in some cases, prefer to ignore it.
- There is a misconception that PP is not science based, since there is a lack of awareness regarding the possibility of measuring PP progress in a relatively objective manner, like the other preparations.
- Coaches are the dominant figure in the training process and may be intimidated by the involvement of the psychological consultant–the "third person" in the training process.
- The sport psychology consultant should understand that he or she is part of the team that helps the athlete to achieve excellence; however, the main figures in the success are the athlete and the coach.
- PP is provided mainly in a classical laboratory setting, without integration within the field work. Most sport psychology consultants misunderstand how to transfer, and how to appropriately and systematically apply, the psychological skills within training and competition.

- The sport psychology consultant may not understand the theory and methodology of sport training within the specific sport discipline. Moreover, in most cases the educational program for sport psychologists does not include knowledge regarding the theory and methodology of sport training and sport disciplines. Therefore, PP is carried out as a separate preparation and does not integrate with other athletic preparations. It is necessary to integrate PP in sport training, and an example of this integration will be discussed extensively in chapter 3.

Chapter 2

FUNDAMENTALS OF SPORTS NUTRITION

James Hoffmann, PhD / Scott Howell, PhD

Although many sport scientists and strength coaches have successfully adopted modern periodization practices to enhance athlete performance, nutritional practices often remain misunderstood and underutilized. Like sport science, sport nutrition has grown and evolved into a developed and well-respected discipline that has permitted coaches and athletes from all walks of life to develop strong evidence-based practices in nutrition. Although still a relatively young science, there is an overwhelming consensus among sport nutritionists regarding most of the fundamental components of a successful sport nutrition intervention. However, pseudoscience and misinterpretation of findings often lead to confusion on how to incorporate nutrition into an effective training plan. This chapter aims to establish the foundational concepts of nutritional periodization, and the practical application of these concepts into an annual training plan.

There is often uncertainty about the role of nutrition due to the sheer scope of uses and applications. Many individuals use nutrition to improve health, physical and mental function, independence, and longevity. In clinical practice, the aim of nutrition is therapeutically to prevent and treat disease, while maintaining or improving health, in a variety of settings. Sport nutrition, however, is brilliantly simple and obstinately seeks one major goal: the enhancement of sport performance through nutritional intervention. This is an important concept to establish early on, as many often confuse nutrition used in general health and clinical practice with that of sport nutrition. That is not to say that sport nutrition does not concern itself with health, but rather the primary **goal** is to *enhance the probability of success for the athlete*. This performance enhancement generally arises from improvements in:

- Metabolism
- Body composition
- Recovery and adaptive processes

DIETARY FACTORS FOR SUCCESS

The areas of sport and exercise science, as well as sport nutrition, are often riddled with fads, misinformation, and biases that can be distracting and misleading to coaches and athletes trying to enhance performance. Although many concepts come and go, wax and wane, or hibernate and reappear, the areas of sport science and sport nutrition have established foundational principles and concepts that are universally applicable, which allow for critical thinking rather than guesswork. Just as training must address the concepts of overload, specificity, phase potentiation, and the other training principles, evidence-based nutrition plans must account for the dietary factors leading to success:

1. Calorie/Energy balance
2. Macronutrients
3. Nutrient timing
4. Food composition
5. Supplementation

These factors allow coaches and sport scientists to stop asking ineffective questions regarding which diets work or if a particular practice is right for an athlete and begin focusing solely on concepts that provide results. In sport nutrition, there is no named diet to follow with trendy or fun quirks, but rather a series of guiding principles that fundamentally drive the athlete to be successful. However, we must keep in mind that not all factors are equally valuable and, in fact, many are distinctly more powerful than others. In order to most effectively implement these factors, they must be rank ordered to allow coaches and athletes to dedicate more time to the things that matter most. Thereafter, fine adjustments are made once the fundamentals have been established.

CALORIE/ENERGY BALANCE

One dietary factor stand outs above all and arguably receives the least amount of attention in the world of sport nutrition. Calorie balance is the most powerful tool in the toolkit when addressing sport performance enhancement through body composition and enhancing recovery and adaptation (Bellisle, McDevitt, & Prentice, 1997; Howell & Kones, 2017; Loucks, 2004; Loucks, Kiens, & Wright, 2011). Calorie balance represents the net energy state of an individual and includes the ratio of total calories consumed and total calories expended through all sources. This can be evaluated on a daily level, or the calories consumed relative to calories expended per day, but also on a longer scale such as a seven-day weekly average. Ultimately calories consumed will be used to fuel normal body functions, the work performed during sport and exercise, and the processes involved in recovery and adaptation. Calorie balance has been estimated to represent a strikingly impressive 50% of relative importance to dietary success and cannot be understated as a critical component of athlete success.

There are three forms of energy balance:

1. Negative (hypocaloric): The state in which the athlete expends more calories that consumed. An energy deficit will result in the breakdown of stored energy sources (carbohydrates, fats, proteins) within the body to compensate for the net loss. As a result, a negative energy balance will yield a loss in bodyweight over time (figure 2.1).

2. Balanced (eucaloric): The state in which the athlete's intake of energy is virtually identical to the energy expended. Although normal day-to-day fluctuations in bodyweight occur, it is virtually impossible to reach a true eucaloric state on a daily basis. The consistency of net energy balance results in a stable bodyweight over time (figure 2.2).

3. Positive (hypercaloric): The state in which the athlete consumes more calories than expended. Excess energy consumed will be used to synthesize new molecules for structure or future storage. As a result, a positive energy balance will yield a gain in bodyweight over time (figure 2.3).

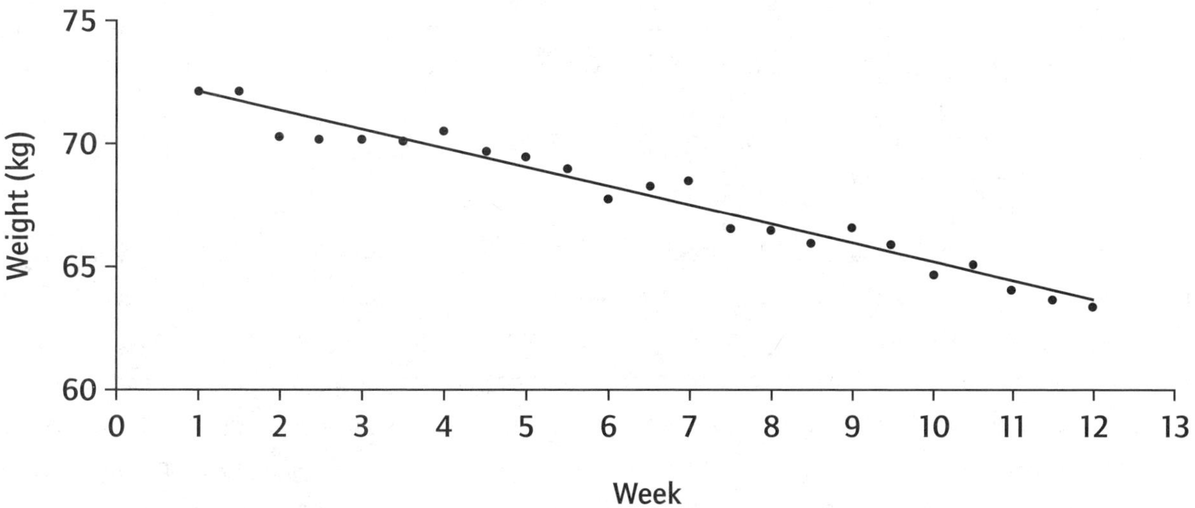

Figure 2.1 Example of a hypocaloric diet

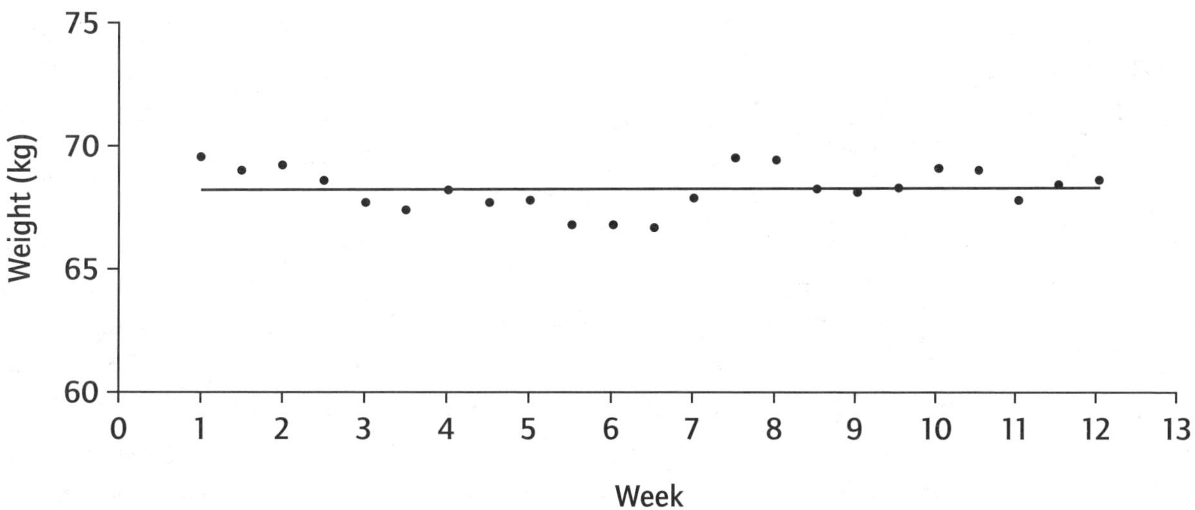

Figure 2.2 Example of a eucaloric diet

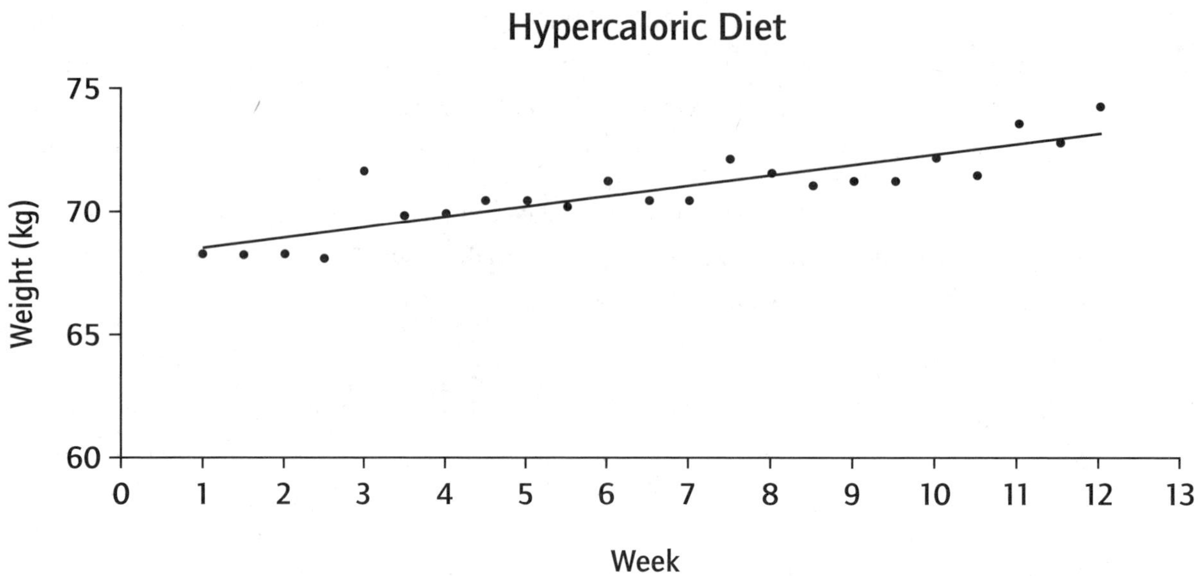

Figure 2.3 Example of a hypercaloric diet

Of all the variables one can account for, energy balance will have the largest impact on an athlete's ability to gain muscle, lose fat, recover from acute exercise, adapt to chronic training, and perform at their best when they need to. Beyond supplying raw materials for energy extraction and molecule manufacture, the state of energy balance has a profound effect on underlying intracellular signaling pathways, enzymatic activity, and hormonal conditions, which regulate all the processes of metabolism. Although many people still try, no one can violate the laws of thermodynamics. A calorie is simply a calorie, which means if an athlete seeks to gain weight, he or she must consume more calories. Likewise, if an athlete seeks to lose weight, the caloric intake must be reduced, energy expenditure increased, or some combination of the two. Controlling energy balance is the first step in designing a successful sport nutrition intervention. But how do we know how many calories are appropriate for any given athlete? How do we know what state he or she is currently in?

There are several useful tests and equations that can help coaches and sport scientists calculate or estimate appropriate daily calories for an individual. Resting metabolic rate tests, the Harris-Benedict equation, and estimated energy expenditure rates, usually per minute, for various activities have long been used to calculate and estimate the caloric needs of athletes. Some of these methods may be more invasive, cost prohibitive, or too complicated for many coaches and athletes to use on a regular basis. One instrument, however, is relatively inexpensive, can be used at will, and provides reliable and practical data regarding an athlete's energy state: the scale. To determine energy, the scale is one of the best tools at your disposal. Simply by tracking bodyweight over time (measured under the same conditions), clear trends in energy balance emerge without the use of statistics, complex equations, or multiple gadgets. If the athlete is gaining weight over time, they are hypercaloric. Simple, easy, and effective. The more common question that occurs, however, is how to determine a balanced energy state for an athlete, or rather the number of calories needed to be eucaloric. This can be done simply by asking the athlete to keep a detailed food log, preferably a baseline of at least two weeks, for comparison with the bodyweight data. Although this is, of course, subject to human (and athlete) error, it remains a practical way of determining this crucial step.

Once food log and bodyweight data have been collected for two weeks, it should be clear the approximate calories on average the athlete is consuming per day as well as the current bodyweight trend. The intent is to approximate a eucaloric state that can be modified as needed to meet specific performance and body composition goals. If an athlete was trying to reduce bodyweight/body fat, consuming 4000 calories a day on average, and bodyweight remains unchanged, a determination of a eucaloric state could be made indirectly on the current activity level, bodyweight, and consumption. To drive weight loss further, the athlete must either increase calorie restriction more, or increase training volume, or both. This process is easier with frequent tracking of weight and intake. Given the increased availability and access to dietary and fitness applications, tracking calories has become significantly easier than ever before. Coaches and sport scientists can easily source these and similar tools to reduce the difficulty and burden on the coaching staff and athletes to determine calorie balance. This allows coaches and practitioners to know the approximate calorie needs of athletes continually and update caloric intake when warranted.

MACRONUTRIENTS

Once the appropriate amount of calories has been determined to match a eucaloric state, the next sequential and proportionately important step is to apportion the calories with the correct proportion of macronutrients, or the amount of carbohydrates, proteins, and fats eaten per day. Although calorie balance is the most important tool in the sport nutrition toolbox, having the right proportions of macronutrients plays a key role to ensure the athlete is gaining or maintaining muscle mass, recovering sufficiently from intense training, and physically prepared for the stresses of training and or competition (Aragon et al., 2017; Beelen, Burke, Gibala, & van Loon, 2010; Burke, Cox, Culmmings, & Desbrow, 2001; Burke, Kiens, & Ivy, 2004; Cermak & van Loon, 2013; Howarth et al., 2010; Morton, McGlory, & Phillips, 2015; Phillips, 2012; Phillips, Moore, & Tang, 2007; Phillips & Van Loon, 2011). Dividing the total daily calories into the correct proportions of macronutrients has been estimated to account for roughly 30% of relative dietary success. Understanding the role of macronutrients in sports nutrition warrants a brief look at the following descriptions.

PROTEIN

Dietary protein quite literally supplies the building blocks of skeletal muscle. It provides the raw materials needed to not only promote protein synthesis and muscle hypertrophy but also reduce protein degradation rates and maintain existing muscle tissue under unfavorable energy conditions. In the context of body composition alteration, dietary protein is the most important macronutrient to be considered. However, within the context of recovery and performance, it moves from the number one spot down to number two, as carbohydrates will have the largest impact on the athlete's ability to recover and perform at a high level.

CARBOHYDRATES

Carbohydrates are the primary energy substrate for virtually all sporting and exercise activities (except for extremely long duration activities) and are the preferred source of fuel for the central nervous system. Carbohydrate is stored within the body as glycogen, which can be found both in the liver where it is used to help regulate blood glucose levels and within the sarcoplasm of the muscle itself, where it is the preferential fuel for high-intensity exercise. Although glycogen can be stored semi-abundantly within the body, it is by no means infinitely sustainable and in some cases, can be a limiting factor in exercise performance. As glycogen levels decrease, the ability to perform exercise or sport tasks become increasingly difficult, often resulting in poor performance in single exercise bouts, sessions, or subsequent sessions. Although a very small amount of carbohydrate can be regenerated through gluconeogenic precursors, there is no magic to replenishing lost energy stores: stored carbohydrate will become depleted during intense or voluminous exercise, which will not return until exogenous carbohydrate is consumed. Although many coaches and nutritionists still seemingly advocate for low-carb dieting in sport, the empirical and anecdotal evidence in favor of dietary carbohydrate is overwhelming. Some have offered the position stance, "You don't need carbs to survive ," and an appropriate reply might be, "You don't need a gold medal to survive either."

The presence (or lack thereof) of carbohydrate also plays a role in regulating the intracellular signaling and hormonal milieu of the cell. The balance of anabolism and catabolism is very sensitive to cellular energy conditions, particularly stored carbohydrate. Chronically low muscle glycogen is not only associated with poor exercise performance, it has also been associated with an increased potential for muscle wasting, hypoglycemia, and blunted adaptive potential from training (Burke et al., 2004; Cermak & van Loon, 2013; Howarth et al., 2010; Zoorob, Parrish, O'Hara, & Kalliny, 2013). Of particular note is the effect of dietary carbohydrate on insulin, one of the primary anabolic hormones of the body. Insulin is secreted by the pancreas and plays a major role in blood glucose regulation. Secretion of insulin is stimulated to a large extent by increases in blood glucose levels as a result of carbohydrate consumption. Insulin acts to facilitate the uptake glucose and amino acids from the blood into their storage forms within skeletal muscle and the liver or potentially as adipose tissue. As a result, insulin plays a large role in the ability to recover from exercise and provide the amino acids needed to grow new muscle tissue. Although carbohydrates may not make up the structure of skeletal muscle, they still play a large role in the ability of the muscle to recover, grow, and resist degradation.

Within the context of improving body composition, carbohydrates are ranked as the second most important macronutrient to consider and control after protein. It mainly serves as an agent to fuel intense and voluminous workouts and provide a powerful anabolic stimulus to the skeletal muscle. In contrast, during performance and recovery, carbohydrates are moved to the most important position and must be the first macronutrient accounted when proportioning daily calories. Substrate depletion is a primary mechanism of fatigue and simply cannot be alleviated by resting or taking time off. If dietary carbohydrate is not consumed, endogenous stores will be chipped away by the ongoing activities of the central nervous system and any subsequent physical activity, at the expense of exercise and sport performance. The power of carbohydrates' effect on the athlete's ability to perform and recover cannot be understated.

FATS

Fats play a role in different body systems for a wide variety of functions. Although fats are required for survival and play significant roles in endocrine function, they have a less profound importance within the role of sport nutrition. Most individuals store more potential energy as fat than they could possibly ever use in a single exercise bout or even multiple exercise bouts. Likewise, dietary fat can't be used as fluidly as carbohydrate for energy production, nor does it aid significantly in favorable body composition change. Within the contexts of body composition alteration as well as recovery and performance, fats are usually the last of the three macronutrients to take into account. However, there are some unique aspects of fat that make it an ideal candidate for altering calorie balance.

In general, fats are highly palatable, or taste good, making them an easy option to add calories to an existing diet plan. Fats also have a higher energy density per gram (9 kcal) than carbohydrate or protein (4 kcal). In addition to being pleasant to eat and providing a large supply of energy, fats such as olive oils, nut butters, avocados, and many others also mix well with other foods and their presence (or lack thereof) can often go unnoticed. Imagine having to add 200 calories from either brown rice (~ 1 cup), or from olive oil (~1.7 tbsp), and ask, "Which one will be more noticeable?" Monounsaturated fats have been shown to be excellent in promoting health, making them an excellent food source to buffer calories once carbohydrate and protein demands are met. Although fats do not have a directly profound role in enhancing sport performance, they do provide an excellent calorie safeguard to meet the daily caloric needs of the athlete.

INTAKE AND TIMING CONSIDERATIONS

PROTEIN

The amount of protein consumed per day is customarily based on the lean body mass of the athlete. In general, athletes with more lean body mass will benefit from higher daily consumption of protein and vice versa. However, for practical purposes, the athlete's bodyweight may be substituted for a measurement of lean body mass, so long as they are within reasonable boundaries of body composition. For athletes who are particularly large or small, specific measurements of body composition may be the best option.

The minimum recommended daily consumption of protein, or the minimal amount of protein needed to grow or maintain muscle mass, is approximately 1.3g per kg of bodyweight per day. Intakes lower than this amount will put the athlete at a higher risk for muscle wasting during hypocaloric conditions, or not gaining as much muscle as they potentially could during hypercaloric conditions.

Optimal daily doses of protein seem to occur around 1.8 – 2.2g per kg of bodyweight per day across a wide variety of sports and activities (Helms, Aragon, & Fitschen, 2014; Houston, 1999; Phillips, 2012; Phillips et al., 2007; Phillips & Van Loon, 2011). Daily intakes of > 2.8g per kg of bodyweight generally do not provide any additional benefit (outside of just additional calories). Intake of protein above and beyond this amount is at the expense of calories from more useful macronutrients such as carbohydrates or fats.

Although there is ongoing and lively discussion and research occurring around timing considerations for protein intake, the current evidence seems to suggest only a small effect, if any, for timing protein ingestion around physical activity. However, evidence also suggests that timing protein intake plays a role in balancing protein synthesis and protein degradation rates throughout the day. At present, it appears that consuming a significant source of dietary protein every 3-5 hours by splitting up meals into smaller and more frequent meals (4-7 per day) tends to have a more beneficial effect on promoting protein synthesis, blunting protein degradation, and enhancing body composition than a more traditional meal structure (1-3 meals per day). For instance, a 63.6kg female sprinter would likely benefit from splitting 140g of daily protein (2.2g per kg bodyweight) into 5 meals of 28g each (140 / 5) spread evenly throughout the day, rather than two meals at 70g each (140 / 2) spread far apart (Aragon & Schoenfeld, 2013; Helms et al., 2014; C. Kerksick et al., 2008; C. M. Kerksick et al., 2017; Zoorob et al., 2013).

There is also evidence to suggest a dose response relationship with protein intake and skeletal muscle protein synthesis. Although there is no clear answer at the current time regarding how much ingested protein per meal can be used toward enhancing skeletal muscle protein synthesis, there does appear to be a theoretical upper limit. This upper limit is highly variable due to several different factors and only specific toward skeletal muscle protein uptake, as dietary protein serves a diverse myriad of physiological systems and body functions. In essence, meeting daily protein needs with infrequent large boluses of protein may be a suboptimal strategy due to this dose response upper limit, whereas having smaller more frequent boluses of protein may allow for a greater relative portion of ingested protein to be taken up by the skeletal muscle.

CARBOHYDRATE

Despite those who still choose to refute the role of carbohydrate in sport nutrition, the evidence is clear: Carbohydrates are the primary energy substrate for virtually all sporting and exercise movements. Accordingly, its dosage is directly proportional to the athlete's total training load from all sources. Table 2.1 details carbohydrate dosage based on activity levels (Beelen et al., 2010; Burke et al., 2001; Burke et al., 2004; Cermak & van Loon, 2013; Shi & Gisolfi, 1998). **Note:** For individuals in excess of 20% body fat, lean body mass may be used in lieu of body weight to perform the calculations.

Table 2.1 Carbohydrate intake by activity level

Activity Level	Recommended Dose Per Day
Minimal (No structured training or sport practice, mostly sedentary throughout the day)	0–1.1 g per kg bodyweight
Light (Planned light days, <60 minutes of hard exercise, light activities of daily living, rehabilitation)	1.1–2.2 g per kg bodyweight
Moderate (Normal training and/or sport practice, up to 90 minutes of hard exercise, very active lifestyle)	2.2–3.3 g per kg bodyweight
Hard (Multiple training and/or practice sessions per day, up to 2.5 hours of hard exercise)	3.3–4.4 g per kg bodyweight
Very Hard (>2.5 hours of hard training per day, endurance sport training, training more than 2 times per day)	4.4–6.6 g per kg bodyweight
High Level Endurance Sport	7.0–10.0 g per kg bodyweight

In principle, the more total training per day, the greater the need of daily intake to fuel and replenish lost carbohydrate stores. Likewise, if training levels are relatively low, the demands for carbohydrate are also low, especially if proper glycogen replenishing protocols are followed post exercise. It is important to note that carbohydrate intake is governed by exercise volume, which often becomes confused with exercise intensity. For example, Crossfitters, sprinters, and weightlifters generally have very intense and difficult training sessions, but the amount of exercise time is actually, relatively low, and more often than not, lower volume loads are produced. Endurance sports on the other hand are commonly performed at relatively low intensities, but the amount of actual exercise time is humongous and higher training loads are produced. Although exercise sessions can be really challenging and strenuous, it's important to remember, for the purposes of carbohydrate intake, that the total duration or volume of exercise determines the dose.

The principal times for carbohydrate ingestion are the periods proximal to training sessions. That does not mean consuming carbohydrates throughout the day is necessarily a bad idea or unbeneficial, but rather the best times to enhance exercise performance, body composition, and recovery are arguably, the pre-, intra-, and post-exercise time periods. Pre-exercise carbohydrate intake has been shown to help manage drops in blood glucose at the onset of exercise, increase carbohydrate oxidation and glycogen breakdown, and dampen fat oxidation and fatty acid mobilization. Carbohydrate during exercise has shown to help maintain blood glucose levels, promote carbohydrate oxidation, spare muscle and liver glycogen stores, facilitate glycogen synthesis during intermittent exercise, and positively influence the nervous system and motor skills. During prolonged exercise, it is recommended to consume carbohydrate at a rate of about 70g per hour. Post-exercise carbohydrate consumption is, perhaps, the most important and has been shown to play a tremendous role in promoting glycogen resynthesis, blunting the catabolic effects of exercise, enhancing protein synthesis (with concurrent ingestion of protein), and setting favorable cellular conditions to promote the RA processes.

Like protein, evidence suggests carbohydrate operates on dose response relationships with similar limitations in terms of how much intake can be taken up and stored as glycogen. Recommendations for optimal glycogen resynthesis range from approximately 0.8 – 1.2g per kg of bodyweight per hour (Jeukendrup & Gleeson, 2010). Most athletes should aim for an intake of roughly 1g per kg of bodyweight per hour post exercise for a duration of one to three hours. On a similar note, the body cannot indefinitely store carbohydrate as glycogen. An average adult stores roughly 300-400g of glycogen in skeletal muscle and 100g in the liver. For example, a 100kg athlete replenishing carbohydrate at a rate of 1g per kg of bodyweight per hour (100g per hour) has the potential to fill lost glycogen stores relatively quickly. To reiterate, this does not suggest eating more carbohydrates throughout the day is incorrect, but rather it illustrates the point that there are potential upper limits to per meal, and per day, carbohydrate consumption.

FATS

Fats, unlike carbohydrates and proteins, do not have explicit dosage or timing considerations. In general, fats can be used as a calorie buffer, meaning they can be added or subtracted to meet energy balance needs after protein and carbohydrates are accounted for without any major limitations. Reducing fats to levels lower than about 0.7g per kg of bodyweight per day can potentially manifest negative side effects in the endocrine system and exacerbate overall fatigue (Burke et al., 2004; Hooper et al., 2012). Short periods of low intake can be tolerated without drastic interference with the training process.

Fats and fiber do tend to slow down digestion and should be avoided during training times to prevent this effect. Although great for satiety and maintaining fullness, high fat or fibrous foods do not possess the favorable effects that intake of carbohydrates and rapidly digesting proteins during training times exert on enhancement of glycogen resynthesis. Therefore, a reasonable approach is to taper fats and fibrous foods throughout the day away from training times. If training is performed early in the day, fats and fiber start low and increase as the day goes on. Likewise, if training is later in the day, fats and fibers start high and taper down toward training times. In the case of having multiple training sessions per day, a prudent strategy would keep fats and fiber low during and between training periods to prevent GI distress and reduction in glycogen resynthesis rates.

NUTRIENT TIMING CAVEATS

On the surface, nutrient timing is a straightforward and rational concept, however, *Caveat lector* directly, applies and warrants further discussion. Currently, it appears that nutrient timing is still worth pursuing, but may not be as powerful as once thought (Aragon & Schoenfeld, 2013; Jeukendrup & Gleeson, 2010; C. Kerksick et al., 2008; C. M. Kerksick et al., 2017). Nutrient timing deals with the timing and frequency of meals in order to optimize their effect on performance, usually corresponding to exercise times. The relative effect of nutrient timing has been estimated to offer 10% of total dietary success for body composition and performance goals. This may not seem profound, but it certainly commands more than a passing glance.

Carbohydrate seems to be more sensitive to the effects of timing, particularly the pre-, intra-, and post-training periods which have been established to exert a positive effect on performance, body composition, and recovery. Protein, on the other hand, has been a subject of controversy, sometimes with conflicting findings or recommendations. Prominent researchers in this area have suggested that there seems to be a frequency consideration for protein intake; however, protein intake proximal to exercise may not have a major effect (Aragon & Schoenfeld, 2013). There may be a positive trend with intake of protein during exercise periods; however, these effects are often not statistically nor substantially different from intake at other times. It has been suggested that this effect may only be relevant to higher level athletes; however, the authors recommend it is likely beneficial to pursue even if the effect is small. Fats appear to have no major effects of nutrient timing outside of the previously mentioned conditions of having a slowing effect on digestion and should generally be lowered around training times.

For athletes who only train once per day, the effect of nutrient timing is small, but worthwhile to optimize and ultimately succeed with their diets. Athletes who train multiple times per day, however, will likely gain a greater benefit from effects of nutrient timing than those who only train once per day. The main reason for this benefit is that glycogen resynthesis can be a major limiting factor to performance when exercise bouts are timed closely together. Resynthesis of glycogen from ingested carbohydrate begins to occur on the scale of hours post consumption and can take upwards of 24 hours to completely refill from voluminous training. Following nutrient timing protocols such as glycogen repletion strategies can help ensure that this process occurs at an optimal rate. Hence, less metabolic fatigue carries over into subsequent exercise bouts.

Finally, confusion persists about the role of nutrient timing and its effect on metabolic rate, specifically, the myth that eating more frequently can cause an increase in metabolic rate. The overwhelming consensus shows this to be incorrect and eating more frequently when matched for total calories does not have any major effects on metabolic rate. One truth is that eating more or less total food can influence the metabolic rate; however, there does appear to be an effect of timing on metabolic rate when daily calories are equated. The takeaway recommendation is that athletes eat about 4-7 meals throughout the day, spaced relatively evenly with carbohydrates emphasized around training times.

Nutrient timing may not carry the same weight as calories and macronutrients, but it is still worthy of pursuit. For recreational athletes or those who train less than once per day, on average, the effect is not substantial. However, for those who train several times per day and are fractions of a percent away from being at the top of their division, nutrient timing has a real and tangible effect.

FOOD COMPOSITION

Investigations into food composition essentially seek to answer whether certain foods are more advantageous that others. Presently, this is an excellent topic for discussion because science paints a clear substantive picture on this issue; however, trends in the fitness industry and popular media depict an image quite the opposite. It is very intriguing how this particular topic has gained so much controversy, yet remains of relatively little consequence in its actual effect. In fact, for the purposes of body composition enhancement and sport performance, the effect of actual food choices you make has been estimated to contribute only about 5% to dietary success. In the context of health promotion, food composition is a much more important factor for success. However, for sport nutrition it remains a relatively small factor. It is reasonable to question if a lean or muscular condition can be achieved on junk food or gain fat eating exclusively healthy foods. The answer is quite simple. Yes, but this is likely, not the best approach for a host of reasons. In sport nutrition practice, the general rule is to emphasize quantities of food rather than qualities of food. However, small improvements can be accrued by making smart food choices throughout the day. As such, a deeper look at nutrient sources and composition is needed.

In general, protein is the most important macronutrient that must be accounted for, since skeletal muscle is literally comprised of dietary protein. Protein is composed of varying chains of amino acids, which may differ significantly across food sources. Accordingly, many different protein quality scores have been developed to evaluate a protein source's "completeness," or how well it serves the amino acid requirements of the human body. One of the most commonly used scores is the Protein Digestibility Corrected Amino Acid Score (PDCAAS), which has been adopted by the Food and Drug Administration and other notable agencies, to evaluate how well a protein source meets the demand of the human body (Schaafsma, 2000). Sources are ranked from 0 – 1.0 with zero being the lowest score and one being the highest. Table 2.2 shows example protein sources and the respective protein quality scores.

Table 2.2 PDCAAS scores of various protein sources

Protein Source	PDCAAS Score (0–1.0)
Casein Protein	1
Whey Protein	1
Soy Protein Isolate	1
Egg White Protein	1
Milk Protein	1
Beef	0.92
Most Fruits	0.76
Most Vegetables	0.73
Kidney Beans	0.68
Peanuts	0.52
Rice	0.5
Whole Wheat	0.42

Although this method of assessment is certainly not perfect, it does provide insight on which protein sources may be the most beneficial. There is a clear trend that milk-based proteins and protein supplements provide some of the highest quality protein sources. These are followed closely by animal proteins, with a declining trend moving into fruits, vegetables, and grains. Thus, for athletes trying to maximize recovery and hypertrophic gains, consuming dietary protein primarily from protein supplements, milk and dairy, and animal-based proteins might provide a slight edge over eating exclusively plant-based proteins. This does not mean that athletes who adopt vegan or vegetarian lifestyles cannot be successful, but rather these athletes should remain more diligent in researching proteins sources and combinations to achieve the same level of completeness that dairy or meat consumption affords.

Another consideration with dietary protein is the rate of digestion. Like the idea of Glycemic Index (GI) for carbohydrates, proteins can also have differing rates of digestion, which in turn may influence how quickly amino acids enter the blood stream and the satiety experienced from the meal. Typically, the various forms of whey protein and egg white proteins tend to be quickly digestible, which makes them an excellent choice around training periods (Graf, Egert, & Heer, 2011; Hayes & Cribb, 2008). However, these sources can leave an athlete feeling very hungry within about 45-60 minutes after consumption when used as a meal, making it a poor choice throughout the day. Animal proteins, casein, and plant proteins tend to be digested much more slowly and have a large satiating effect, making them an ideal choice throughout the day away from training times. For the most part, protein sources should be primarily lean (≥90%) with reasonable fat content. Protein with high fat content, such as bacon, sausage, whole milk and creams, processed meats, or fatty pork or beef products, should be avoided. Sources like chicken, turkey, fish, lean beef and pork, dairy, eggs and egg whites, and protein supplements need to be prioritized over high fat content sources.

Carbohydrates, in the scope of food composition, follow similarly to proteins; in other words, the type of carbohydrate you choose in the big picture is largely unimportant for sport nutrition. However, the choice of carbohydrates often entails largely practical considerations for implementation. Most sport nutrition resources, including the authors of this book, recommend that the majority of an athlete's carbohydrate intake come from food sources including whole grains, whole wheats, fruits, vegetables, and other wholesome items. These types of carbohydrates are generally high in fiber and micronutrients, low glycemic, and have a large satiating effect. Regardless of these important characteristics, for many athletes who are eating well above their bodyweight in carbohydrates per day, this is simply not achievable by exclusive reliance on these sources. For example, a serving of brown rice or oatmeal at ~ 100g of carbohydrate can be a daunting task even for experienced athletes, not to mention additional macronutrient content when eaten as a meal. On the other hand, 100g of carbohydrate from sports drinks, fruit snacks, or kid's cereal can go down quickly and easily.

In order to strike a balance between making wholesome food choices and ease of implementation of a diet program, coaches and athletes can use the GI of carbohydrate sources to their advantage for a number of reasons. Recent evidence suggests the actual GI of carbohydrate is relatively unimportant in enhancing performance, with the total quantity of carbohydrate being the most important factor (Mondazzi & Arcelli, 2009; O'Reilly, Wong, & Chen, 2010). Manipulating the GI of meals is likely more important for athletes who train multiple times per day, where the rate of glycogen resynthesis can be a limiting factor to performance, and much less important to those who train once per day or less. Regardless of these points, the GI can be used as a heuristic to guide selection, not a steadfast mandate. In general, high GI foods tend to enter the blood stream very quickly, have a more profound insulin response, and are very easy to consume in large quantities. These foods are great choices during the intra workout and immediate post-workout periods to help enhance glycogen resynthesis, exogenous glucose oxidation, and managing blood glucose levels during exercise. Unfortunately, these are not prime choices during non-workout periods due to the low satiety effect, tendency to increase food cravings, and general lack of fiber and micronutrients. Lower GI foods make for excellent choices during non-workout periods due to the greater satiety effect, high fiber and micronutrient content, and reduced insulin response. Examples of GI in different foods sources are shown in table 2.3.

Table 2.3 Glycemic Indices of various foods

Carbohydrate Source	Glycemic Index
Dried Dates	103
Glucose	99
Lucozade	95
Cornflakes	81
Gatorade	78
Glazed Doughnut	76
White Bread	73
White Rice	64
Potato Chips	54
Wheat Bread	53
Orange Juice	50
Low Fat Ice cream	50
Carrots	47
Orange	42
Apple	38
Bran Cereal	38
Whole Wheat Spaghetti	37
Skim Milk	32
Kidney Beans	28
Low Fat Yogurt	24
Peanuts	14

For relatively new or recreational athletes, manipulating the GI is often not necessary, and the best approach is to adhere to traditional recommendations of whole grains and wheats, fruits, and vegetables when this can be practically implemented. These sources of carbohydrates tend to keep athletes full throughout the day and serve to promote general health. More developed athletes or those who train multiple times per day may see a slight benefit in focusing on higher GI foods around training times to enhance recovery and tapering down the GI during non-training times to remain satiated and stabilize blood glucose.

It is worth noting for all groups that although the GI does not appear to have a significant effect when calories and total carbohydrates are controlled, there is emerging evidence that suggests consuming high sugar and tasty foods has an unfortunate effect of increasing desire for such foods and overall food cravings. Bland and heavily satiating foods appear to control hunger more effectively and possibly even deter food cravings. This might seem superfluous at first but consider an athlete who is cutting weight for three months to achieve an optimal competition bodyweight. Although eating high or low GI foods may not have a tangible effect on performance, overconsumption of junk food could wreak havoc on mental preparedness through incessant food cravings and possibly adherence to the program. Likewise, an athlete that must move up a weight class might experience a road block through consuming copious amounts of brown rice and broccoli to gain 5kg. Although manipulating the GI may not have a direct tangible performance effect, these examples show there are many practical reasons why manipulating the GI can be used to keep athletes on the path to success.

The role of fats in health and disease has been debated for many years. In sports nutrition, research remains controversial on the capacity of fats to confer marked performance benefit. The authors believe fats have an arguably minimal impact on performance in most sports, given basic requirements are met. General trends for health appear to have the most utility athletics.

When choosing fat sources, it would appear that monounsaturated fats, such as those found in nuts, natural nut butters, olive oil, and avocados, seem to have the most positive effect on promoting health. The research on saturated fats, such as those found in butter, cheese, and fatty meats, remains inconsistent. The consensus appears to indicate that in moderation these foods are safe for consumption; however, when eaten to excess or in combination with being overweight or over fat, may have the potential for increasing cardiovascular disease risk. Although more can be said about the role of dietary fat from a health promotion standpoint, it quickly deviates outside of the realm of sport nutrition and should be discussed with a registered dietician or medical professional. Athletes who are also cognizant of health promotion should probably consume the majority of their dietary fats from monounsaturated fat sources, however, the actual effect on sport performance is likely minimal at best.

SUPPLEMENTS

Supplements are last in our discussion of dietary factors for success, frankly because they are the least important. Supplements tend to be the starting point for many athletes and people who are seeking to improve their nutrition; however, they are of minimal importance when compared to the effects of calorie balance and macronutrients (Manore, 2012). Supplements have been estimated to contribute less than about 5% to total dietary success in body composition enhancement and performance. As the name implies, supplements should be...well, supplemental to a well-established nutrition program. This does not imply that they have no value at all, but rather it illustrates the fact that they exert a minimal magnitude of effect.

You may wonder if there are supplements that are worthwhile that potentially possess a performance enhancing effect. The answer is absolutely. Although typical literature reviews on supplements are extensive and vast, the authors of this book are only willing to recommend a few which carry not only an abundance of scientific evidence but also maintain a high degree of practicality as well. Those supplements include:

1. Protein supplements
2. Carbohydrate supplements

3. Stimulants (Caffeine)
4. Creatine monohydrate
5. Beta alanine

This list is certainly not all inclusive and is subject to change over time. The intent is to simply highlight the more efficacious supplements based on our current understanding. What coaches and athletes should understand is that sport supplements should not be the first stop on the nutrition journey, but rather they could possibly be used to fine tune after the fundamentals have been established. Once an individual has grasped and applied the concepts of calorie balance and macronutrients, supplements can be used in conjunction with good nutrient timing and food composition practices to enhance their effects.

ADDITIONAL DIETARY CONSIDERATIONS

HYDRATION

The understanding that hydration is important for maintaining both health and performance is common knowledge for most people. Even a relatively small loss of about 2% in bodyweight can manifest into performance detriments, thus maintaining adequate hydration before, during, and after exercise periods is essential to maintaining high levels of performance (Casa et al., 2000; Jeukendrup & Gleeson, 2010; Judelson et al., 2007; Maughan & Shirreffs, 2010; Shi & Gisolfi, 1998). Coaches and athletes should consistently take steps to maintain euhydration with the awareness that there is no secondary beneficial tier of supreme hydration beyond that, meaning more is not necessarily better. In some cases, drinking water in excess can lead to hyponatremia, particularly in instances when high volumes of water are consumed without additional food or electrolytes, and when the person has been profusely sweating.

For most athletes, a daily intake of about 5 ounces (141.75g) per 100 calories consumed is a good place to start. This, of course, can vary depending on environmental and exercise conditions such exercising in the heat or exercise for prolonged durations. Hydration can generally be monitored through urinary specific gravity, bodyweight changes due to fluid loss from exercise, and monitoring urine color. Although urinary-specific gravity may require the use of a refractometer and the unpleasant exchange of urine, values of between 1.003 – 1.030 are generally an index of euhydration, with greater than 1.030 generally representing a significant degree of dehydration. Fluid lost during exercise can be monitored by changes in bodyweight and should generally be replenished by a factor of approximately 1.5 times the fluid lost. For every kilogram of bodyweight lost, it would be replaced by 1.5 kilograms of fluid (Case et al., 2000; Maughan & Shirreffs, 2010; Shi & Gisolfi, 1998; Zoorob et al., 2013). Urine color is a qualitative form of monitoring, but in general, if urine remains clear to a light yellow color, the athlete is likely sufficiently hydrated. It may be more important for new or developing athletes to maintain a more regimented hydration schedule, whereas more experienced and trained athletes can generally rely more on thirst mechanisms but remain aware and take steps to remain euhydrated regardless of thirst.

MICRONUTRIENTS

Vitamins and minerals play another important role in promoting both health and performance. There are numerous available resources to coaches, athletes, and scientists which describe how each functions within the body, dosing considerations, and the food sources in which they can be found. Our discussion on micronutrients in this book is actually relatively brief simply because the gross majority of sport nutrition occurs on a macro scale, and although micronutrients are worthy of discussion, the take-home message to coaches and athletes is generally the same outside of some very specific circumstances. Typically diets that are rich in lean meats, whole grains, whole wheats, fruits, vegetables, nuts, and oils cover all if not the preponderance of micronutrient requirements. The requirements not being met by diet alone can often be remedied simply with an over-the-counter vitamin supplement. Outside these conditions, our discussion deviates beyond the realm of sport nutrition and into clinical nutrition, which generally should be addressed by a registered dietician or medical staff.

NUTRITIONAL PERIODIZATION

Although a well-developed nutrition plan may address all the dietary factors for success to some degree, at any given time, there are also periods throughout the training cycle where some factors may carry more importance than others. Likewise emphasizing certain components at inappropriate times may lead to suboptimal training or performance. In fact, just as training must be periodized over time to promote short- and long-term athlete development, fatigue management, and physical and psychological preparedness, so too must the dietary intervention to match the needs of the training and competition cycle. Therefore, when integrating nutritional interventions into an annual training plan, we recommend considering the following phases:

1. Body composition alteration
2. Stabilization
3. Recovery and performance
4. Active rest

BODY COMPOSITION ALTERATION

Enhancements in body composition are not limited to weight class sports or activities such as wrestling, weightlifting, and bodybuilding. In fact, it is imperative for all athletes to strike a balance of muscularity, body fat, and bodyweight to compete at an optimal size for success. For any sport, there are generally normative range standards of body composition, which can be a useful tool and provide insight into how the individual being trained matches up against an ecologically valid group. This might seem obvious, but think about the last time you saw a 100m hurdler with a big gut, or an NFL lineman or rugby prop that weighed 75kg. Body composition alterations are not simply for the "meat-head" crowd or those simply trying to be more muscular, rather it is about optimizing an athlete's bodyweight and composition for performance. Portions of the annual plan can then be dedicated to the enhancement of body composition to meet these standards or the optimal competition size of the individual.

What exactly do enhancements in body composition refer to? Generally, this will come in the form of either increasing the lean body mass or reducing the body fat of athletes. With very few exceptions (e.g., sumo wrestling, American football), the pursuit of gaining weight as ballast is generally not an optimal strategy to enhance performance, as more often than not it appears that leaner athletes generally outperform their heavier counterparts. One area of confusion with body composition enhancement has been the pursuit gains in lean body mass while decreasing or maintaining current body fat at the same time, sometimes referred to as "lean gains." Although this phenomenon can occur where an individual concurrently loses fat and gains muscle, it generally only happens under three conditions:

1. The individual is new to structured training.
2. The individual is new to structured dieting.
3. The individual begins using (or uses more powerful) anabolic agents.

This effect is generally short-lived, and over time the gains in muscle or loses in fat tend to slow down to a grinding halt, making the process unnecessarily arduous and time consuming. The amount of fat loss or muscle gained over the course of a year or two could simply have been accomplished in a few months of planned dieting. Thus, for trained individuals, the most efficient use of time appears to be focused allocated periods of either muscle gain or fat reduction. Although this may seem very straightforward, there are many considerations when pairing a body composition-oriented nutrition plan with a training program.

Body composition changes also follow a phasic pattern of alternating sequences. Within body composition changes we must also consider three subphases:

1. Hypercaloric (known as the "mass" phase). Maintenance level calories are increased by about 500-1000 kcal per day which results in weight gain (figure 2.3).
2. Isocaloric (known as the "mid" or "maintenance" phase). Daily energy intake is balanced with daily energy expenditure, which results in a (mostly) stable bodyweight over time (figure 2.2).
3. Hypocaloric (known as the "cut" phase). Maintenance level calories are decreased by about 500-1000 kcal per day which results in weight loss (figure 2.1).

HYPERCALORIC

The goal of the hypercaloric phase is simply to gain weight, with as much of that weight coming from lean body mass as possible. As mentioned previously, it becomes increasingly difficult over time to continue gaining muscle without significantly altering body fat, and it is very probable for most athletes to gain some body fat while pursuing gains in muscle mass. It cannot be understated, however, that muscle grows the fastest on a hypercaloric diet. This makes it imperative to establish a positive relationship with the athlete in terms of the benefits of gaining muscle and their personal performance, the amount of fat they may gain in the process, and the steps involved in reducing fat and returning them to their optimal body composition for competition.

Evidence suggests that the leaner an individual is before starting a weight gain phase, the higher the relative proportion of muscle mass that is gained. With this in mind, it is generally recommended that males with body fat levels of ≥ 15% and females with body fat levels of ≥ 25% not initiate weight gain phases until they have brought their body composition under these values, as a larger proportion of the weight gained will come from body fat (Forbes, 1987, 2000). Gains in bodyweight of about 0.5-2% per week seem to be well supported for most normal sized individuals (Helms et al., 2014; Morton et al., 2015; Slater & Phillips, 2011). Increases in bodyweight beyond this amount will generally result in an unfavorable ratio of muscle-to-fat gain, which will result in longer cutting phases to lose that weight later on. Decreases in bodyweight lower than this amount appears simply to be unnecessarily slow, with more muscle gain possible per the same amount of time.

Because high volume loads are required to gain muscle mass, hypercaloric phases should be mainly included in the general portion of the preparatory phase. Because the general preparatory phase emphasizes developing work capacity, usually with the highest volume loads of the training cycle, it provides an excellent opportunity to make gains in muscle mass. Trying to achieve a caloric surplus or muscle gain during other phases of the annual plan will usually prove less fruitful, as high-volume training will interfere with the ability to express power and explosiveness, as well as sport performance, and eating a caloric surplus without a sufficient volume stimulus will likely result in excess amounts of fat gain.

ISOCALORIC

The maintenance phase is a crucial and underappreciated step in body composition alteration, and it will also play a crucial role in our discussion of the stabilization phase of the annual plan. The maintenance phase serves three main purposes within the body composition alteration phase:

1. Alleviate the physical and psychological fatigue of long-term weight gain/weight loss dieting
2. Stabilize the bodyweight and metabolic activity
3. Allow the body's natural set point time to adjust to the newly achieved body mass and body composition

Neglecting to account for the delayed effects of the body's homeostatic set point is a common mistake when dealing with substantial bodyweight changes. When weight loss occurs through structured and difficult dieting, it is not unusual for an individual to rebound back up to their baseline weight when diet is not controlled (Cornier, 2011; Elfhag & Rössner, 2005; Keesey & Hirvonen, 1997; Klem, Wing, McGuire, Seagle, & Hill, 1998; Sumithran & Proietto, 2013; Weinsier et al., 2000). Similarly, after weight gain phases, individuals are often eager to lose the fat they gained and move quickly into weight-loss phases. This typically results in a quick downward rebound and a reduced net gain of muscle. The maintenance phase is used to bridge the gap between weight gain and weight loss phases or as an opportunity to rest and reset the body during long-term weight loss or gain.

As the name implies, the maintenance plan is a sustainable plan, whereas the mass and cut phases are not. The maintenance phase is a return to normal balanced eating and training habits, and an opportunity to manage physical and psychological fatigue. Maintenance phases are implemented typically after prolonged weight gain or weight loss such as the general preparatory phase, as well as when the athlete is beginning to transition into more sport specific training, such as the specific preparatory phase. It's imperative that the athlete is given time to train at the newly attained bodyweight and body composition, as many sport skills and techniques may be altered in the process. For example, a powerlifter who is used to bracing against a weight belt may have to modify his or her technique after losing 20 lb (9 kg) to make weight. Similarly, a rugby player may need time to adjust and acclimatize the sprinting, jumping, and scrummaging technique after gaining a substantial amount of size and strength. Major body composition alterations should always be stabilized with a maintenance phase for one to three months (depending on the time available) prior to competition. This also provides the secondary benefit of allowing the homeostatic set point time to adjust and allow the athlete to be free of structured dieting concerns leading into competition.

HYPOCALORIC

The goal of the hypocaloric phase is simply to lose weight with an overarching emphasis on reducing body fat rather than lean body mass. Training influences, aside from calories, are potent stimulators of muscle growth and accordingly when they are constricted, muscle gain becomes less and less probable, and muscle loss becoming more and more probable. Additionally, restricting calories will also have a potent effect on stimulating fatigue through substrate depletion, so care and close monitoring should be undertaken when implementing a cut phase. With the correct balance of training, calories, and macronutrient intake, muscle loss can be minimized, and body fat can be reduced over time.

Generally accepted values of weight loss for normal sized individuals range from about 0.5-2% of bodyweight per week (Hall, 2008; Stiegler & Cunliffe, 2006; Westerterp-Plantenga, Nieuwenhuizen, Tome, Soenen, & Westerterp, 2009). Generally, more than this amount will come at an unfavorable cost of fatigue and a greater potential for losing lean tissue. Less than this amount is generally unproductive, and more weight could be lost per the same amount of time. Sport scientists and coaches should also keep in mind that increases of fatigue should be monitored and managed, as too much fatigue will lead to sub-optimal training, practice, and recovery, all of which can lead to reduced adaptive potential.

Similar to the hypercaloric phase, the best time to implement a hypocaloric phase will be during the general preparatory phase. Because of the high-volume loads, achieving a negative calorie balance is more easily accomplished through diet and exercise, and these high training loads will also promote a greater degree of muscle sparing.

Implementing a hypocaloric phase at any other time, however, is highly discouraged, as the amount of fatigue it generates will inhibit the expression of power and explosiveness as well as the ability to execute sport techniques and tactics. Although cutting weight all the way into a competition is a common practice in many sports, it should be avoided whenever possible due to the suboptimal training conditions leading into the event, resulting in less probability for success.

BODY COMPOSITION PERIODIZATION

Within the body composition alteration phase, different combinations of weight gain, loss, and maintenance can be used to achieve specific desired goals. When the preparatory phase as a whole is long, major bodyweight changes and compositional changes for the athlete may occur. When the preparatory phase is relatively short, the athlete will likely be limited to gaining or losing weight for that period of time and adjusting the next training cycle accordingly.

Contrary to popular belief, neither the mass phase nor the cut phase, are sustainable practices over long periods of time. Gaining muscle requires high volume loads and sustained efforts to consume food, all of which are physically and psychologically taxing on the athlete. Likewise, with losing fat, high training loads under caloric deficits are required, all of which is fatiguing. Accordingly, for both weight gain and weight loss, the processes begin to slow down over time and unfavorable physiological and psychological adaptations begin making the processes increasingly difficult. Changes in insulin sensitivity, resting metabolic rate, the desire to train, desire to eat, and the body's edict to return to homeostasis all begin to weigh down the athlete. Because of this, it is generally recommended not to exceed three months of dedicated weight gain or loss without interspersing periods of maintenance ranging from about one to three months. This allows unfavorable metabolic changes to normalize, the bodies homeostatic sets points to catch up and helps alleviate prolonged psychological fatigue from dieting.

For short-term weight loss and weight gain, hypercaloric and hypocaloric phases can be implemented in the general preparatory phase for one to three months at a time, followed by an isocaloric phase lasting one to three months or more depending on the proximity to the competitive phase. Additional weight loss or gain will then be achieved in the next cycle of training when the athlete starts the next general preparatory phase. See figures 2.4 and 2.5 for examples of short-term weight gain and weight loss.

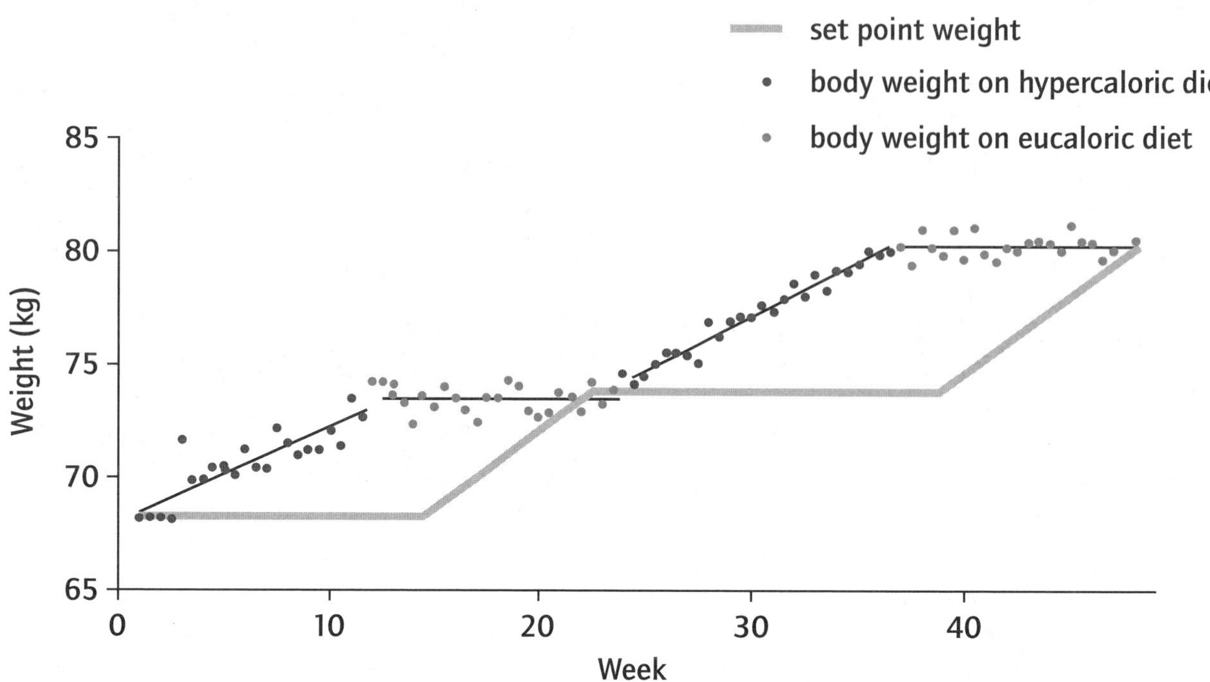

Figure 2.4 Short-term weight gain

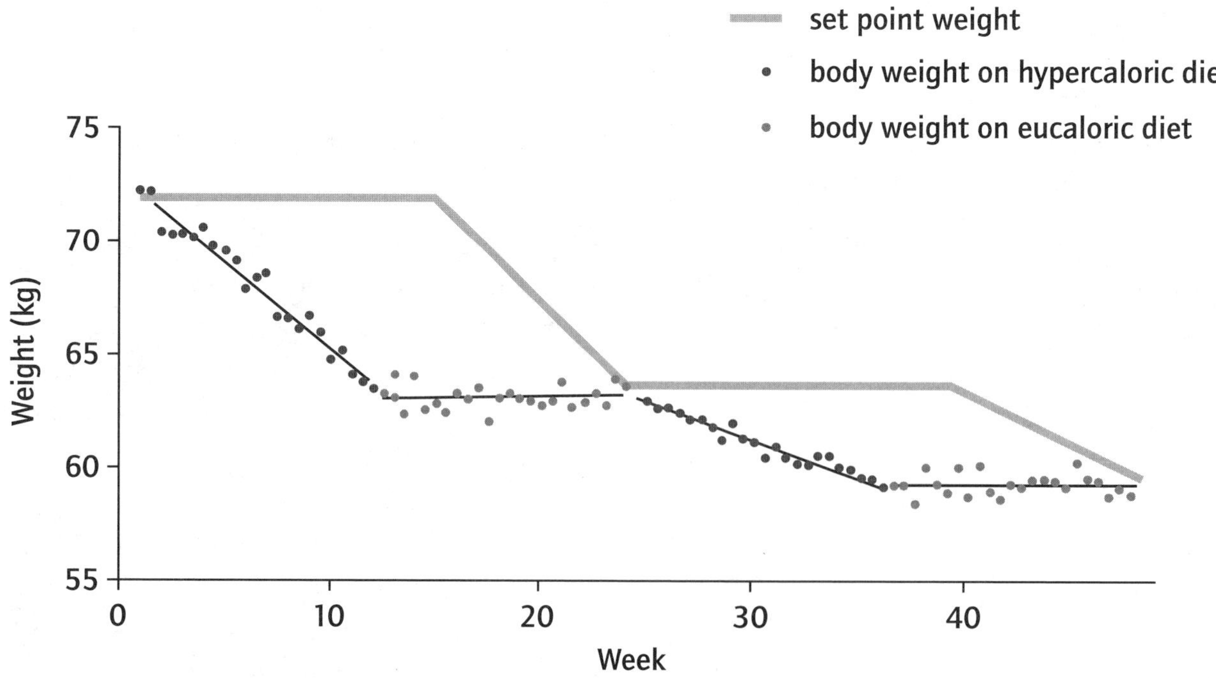

Figure 2.5 Short-term weight loss

For long-term weight loss or weight gain, some slightly modified approaches should be adopted. For long-term weight gain, hypercaloric phases lasting upwards of three months can be implemented one at a time followed by one to three months of maintenance. Once the maintenance phase is complete, the athlete can move into the cut phase to lose the accumulated fat and return over the next three months to the previous baseline body composition, now at a heavier weight. Once the baseline body composition has been attained, the cycle repeats and weight gain occurs slowly through periods of muscle gain and fat reduction.

Coaches and sport scientists should also be aware of the potentiating effects of periodizing phases of weight loss and gain over time. Weight (muscle) gain over time is typically associated increases in resting metabolic rate, decreases in insulin sensitivity, and decreases in metabolic efficiency. Increases in resting metabolic rate and decreases in metabolic efficiency may potentiate subsequent hypocaloric phases and allow for weight loss to occur more easily or rapidly. Weight loss over time is typically associated with decreases in resting metabolic rate, increase in insulin sensitivity, and decreased, increased metabolic efficiency. Decreases in resting metabolic rate and increases in insulin sensitivity can also have a potentiating effect on subsequent weight gain. Hence, with long-term strategies, it appears that mass phases can potentiate weight loss in subsequent cut phases, and likewise cut phases can potentiate weight gain in subsequent mass phases.

For long-term weight loss, hypocaloric phases lasting upwards of three months can be implemented followed by about one to three months of maintenance. Once the body weight has stabilized and the athlete can demonstrate a healthy relationship with dieting and training, additional hypocaloric phases may be implemented, and the process can be repeated. Keep in mind that people tolerate weight loss differently, and the length of the maintenance phases during long-term weight loss should be reflective not only of their physical abilities to continue but their psychological state as well.

Within the body composition alteration phase, many cycles of mass, cuts, and maintenance phases can be sequenced depending on the length of the preparatory phase as a whole and more importantly the general preparatory phase. As body compositional changes occur throughout the general preparatory phase, there will be a distinct paradigm shift as the athlete transitions into the specific preparatory phase and body compositional changes are ceased, training specificity increases, and the athlete moves into the stabilization phase of nutritional periodization.

STABILIZATION PHASE

The stabilization phase is a critical and often overlooked concept in nutritional periodization. As the name implies, the main focus of this phase is setting the approximate baseline bodyweight which the athlete will be competing at or holding before future body composition changes. This is generally achieved by progressively adding or subtracting about 500-1000 kcal per day depending on the previous bodyweight goals, and then holding the target bodyweight for about one to three months. Although this may seem insignificant, allowing the athlete's bodyweight to normalize after a period of significant weight loss or gain can have a substantial impact on the body's set points for body composition and bodyweight, the amount of muscle gained or lost, physical and psychological fatigue management, and normalizing any unfavorable metabolic disturbances.

The stabilization phase is integrated with the specific preparatory phase within the annual plan. Within this phase, there is a marked shift from more generalized training to predominantly sport-specific training. In addition to developing technical mastery of sport and training skills, the specific preparatory phase is used to enhance fitness characteristics such as maximal strength, power, speed, and sport-specific endurance. In order to develop these skills and abilities, the training volume is generally reduced by upwards of 40% of that from the general preparatory phase, as fatigue usually inhibits their expression. Concurrently, volume from sport practice tends to increase as athletes begin spending more time practicing and performing live simulation. Accordingly, the athlete's state of energy balance may change from that of the previous training cycles and may need to be reevaluated in order to stabilize bodyweight.

Many fitness characteristics and sport skills can also be altered by significant changes in body composition. Athletes who gain muscle mass will often find themselves stronger and more powerful than previously before, and likewise those who have lost weight may find their ability to translate strength into power, speed, and explosiveness enhanced from having to propel less bodyweight in movement. Cyclists, for example, may find themselves at an enhanced power-to-weight ratio, with concurrent improvements in economy leading to an altered race pace. There are many examples of how major changes in bodyweight and body composition can affect sport and training performance, for better or worse. Coaches and sport scientists should be aware that many of these changes will likely have an adjustment or acclimatization period for athletes. As the specificity of training increases toward competition, so too must the feel of competition for the athlete. The bodyweight should be stabilizing very close to the competition weight so the athlete is aware of any potential physical, technical, or tactical adjustments that have resulted, and the athlete is able to feel out the pacing and effort needed for competition.

For most sports and activities, it is recommended to stabilize their bodyweight within ± 2% of their competition bodyweight for about one to three months prior to moving into the competition phase. Although it is commonplace in many weight class sports such as weightlifting or combat sports such as wrestling to diet all the way up to the competition, this practice really neglects addressing the quality of life and training leading into the competition. Substrate fatigue from hypocaloric dieting can have a direct negative impact the specialized outcomes of the specific preparatory phase and thus reduce the adaptive potential for athlete and probability of their success. In addition to the decreased adaptive potential for fitness characteristics, fatigue can also have an inhibitory effect on the expression of technique, decision making, and tactics. When combined, the effect of substrate level fatigue at inappropriate times leaves athletes in a state of suboptimal fitness and lessened ability to perform. Lastly, the stress of dieting can be an unnecessary distraction, and as athletes move into the highly specialized portions of their training plan, they should be leaving concerns of bodyweight behind and focusing on training to win. An MMA fighter who chooses to be within 5 lb (2.27 kg) of their weight class one to two months out from competition will likely have a much more productive training cycle than a fighter who needs to drop 20-30 lb (9-13.61 kg) in that same amount of time.

Diets designed for body composition alterations on average tend to be a bit more rigid in structure and unsustainable over long periods of time. As the athlete moves into the stabilization phase, so too must their ability to sustain eating habits that will lead them to success. After completing the body composition alteration phase, the athlete will likely have developed an eating plan or routine with consistent habits and patterns. The stabilization phase seeks to build upon the already established routine by adding or removing calories (and macronutrients accordingly) and alleviating the stress of dieting by allowing more freedom within the areas of nutrient timing and food composition. Here the athlete and coaching staff should still seek to address all the major factors for dietary success, now with foods, meal frequencies, and meal sizes that are not distracting or stressful to the athlete. The athlete who was stuffing himself with high-calorie junk food and feeling the ill effects of having a hard time gaining weight can focus on foods and meal sizes that make him feel better and less bloated. Likewise, the athlete who followed a strict weight loss plan, whose will is crushed and heart rate raises 10 beats per minute at the sight of chicken, brown rice, and broccoli, can enjoy some treats and switch over to foods they enjoy more, while still operating within the boundaries of calorie balance and macronutrients. Luckily most of these habits should be well rehearsed at this point and any adjustments during this phase are for fine-tuning.

The stabilization phase is a resetting period for the athlete. As bodyweight moves into maintenance, training becomes highly specific, and there is a shift from training to improve fitness to training for competition. The stress of dieting and training is reduced, and the body is given the time it needs to adjust to substantial changes in bodyweight and composition. As the athlete moves into the competitive portion of the training plan, the intensity of practice and competition may result in additional challenges in the ability to effectively recover from intense exercise and perform in subsequent bouts. Accordingly, the dietary focus must also shift to address these demands and ensure the athlete is performing at their best even in challenging conditions.

RECOVERY AND PERFORMANCE PHASE

The recovery and performance phase is the period in which the athlete is actively competing, whether it's seasonal play, a series of fights, or a specific event like a triathlon, this is the phase that is used to complement the intense demand of highly specific sport practice and competition. For many athletes this period may include activities such as scrimmaging, running/cycling/swimming at race pace, sparring, or jumping/sprinting/throwing using competition-like conditions and levels of effort. In addition to microtrauma and unfavorable changes in neuroendocrine conditions, substrate depletion, particularly carbohydrate depletion, is a major contributing factor to fatigue and can also lead to suboptimal performance, recovery, and training. Not surprisingly, a carbohydrate-rich diet is one of the easiest and most effective protective mechanisms against fatigue during all phases of training, but of particular importance during the recovery and performance phase.

Unlike the body composition alteration phase, where protein was ranked first above carbohydrates and fats respectively, we make a distinctive shift in priorities during the recovery and performance phase to prioritize carbohydrate intake and ensure the athlete is always resupplying lost glycogen stores. During this phase, calories from fats can be reduced or minimized and substituted with additional carbohydrates beyond the standard recommendations. Protein can also be lowered if necessary, but carbohydrate substitutions should largely come from fats first and proteins on a needs-only basis. Does this mean that protein is no longer important? No, of course not; however when push comes to shove, carbohydrate intake must be optimized before all else, and compromises can be made to dietary fats first, and protein if necessary. In fact, macronutrient priorities are not the only major considerations when transitioning into this phase, as important changes in nutrient timing, food composition, and supplementation may provide additional assistance.

As always, calorie balance is the largest determining factor for success, regardless of phase. As previously mentioned, macronutrient quantities are also paramount, with an increased emphasis on carbohydrate intake. Nutrient timing now becomes an increasingly important factor to consider, as glycogen repletion is often a rate-limiting step for performance, especially when competitions are within the same day or within multiple subsequent days. The ability to replenish muscle and liver glycogen, as well as the ability to use carbohydrate effectively during exercise, can all be influenced by the timing of nutrients. Although questions still remain about the efficacy and effect sizes of nutrient timing practices, the overwhelming consensus seems to indicate a positive effect for optimizing carbohydrate intake within the workout window (pre-, intra-, and post-exercise) on glycogen repletion. For athletes looking to compete at their best, considerations should be made to address nutrient timing factors around training and performance times. This can also include modifying the glycemic index (GI) of the carbohydrates ingested to ensure the most efficient rate of glycogen repletion.

One common misconception in sport nutrition is that virtually all of the carbohydrate consumed should be of high nutrient quality, low GI sources such as fruits, vegetables, whole wheats and whole grains, and so on. At face value, this recommendation is very reasonable and can also help promote a number of additional health-related benefits to the athlete; however, one must also consider the practical implications of this strategy and its potential limitations. For many athletes during the competition phase, it is not unusual to be eating well over two times their own bodyweight in grams of carbohydrate per day, and for sports like Crossfit, cycling, and other endurance sports, it can be even higher. At this point, the question becomes, how much brown rice, oranges, and oatmeal must one eat to reach 500 g or more of carbohydrate per day? Most who attempt this quickly realize that the task can be quite daunting, as many of these foods are voluminous and fatiguing to eat in high quantities. Additionally, too much may result in unwanted GI distress. Many other carbohydrate-rich foods, although not deemed as traditionally healthy foods, can provide a large bolus of carbohydrate to the athlete without feeling overly full. High-to-moderate GI foods such as sports drinks, soda, low-fat cereal, low-fat gummy snacks, and many others are extremely palatable and easy to consume. When consumed around training times they also have the additional beneficial effects of enhanced glycogen uptake from nutrient timing, making them an excellent choice when the athlete needs to refill large energy stores quickly. We must remember that sport nutrition is not health nutrition, and although promoting the health of the athlete is still important, it should not come at the expense of strategies meant to enhance performance. This is bolstered by the fact that food composition makes up a relatively small portion of dietary success and can be modified during the competition phase to ensure the athlete is consuming enough carbohydrate without being constantly overfull or dreading their next meals. In many cases having a post-match blowout of soda, ice cream, and pasta may be a better strategy than simply *not eating enough* before the next competition.

Within the recovery and performance phase, we have determined that calorie balance and macronutrients are still our primary concerns, however, we have placed a greater emphasis on nutrient timing to ensure that the athlete is recovering as quickly and rapidly as possible. In order to do so, we are also concerning ourselves less and less with eating strictly traditionally healthy foods and allowing for flexibility in food composition for practical implementation purposes. It is also worth mentioning supplementation with ergogenic aids during this phase as they may provide an additional benefit, albeit a small one. It cannot be understated that supplementation makes up a very small portion of dietary success and really should only be addressed after all of the other dietary factors are taken into account. Coaches, athletes, nutritionists, and sport scientists alike must all also be aware of the guidelines and restrictions of performance-enhancing substances within their governing bodies. Buyers should always do their research before purchasing ergogenic aids as many companies have not been forthcoming about their actual ingredients, often coming at the expense of the athletes taking them. That being said, another effective dietary strategy during this phase can be the use of creatine monohydrate, which has been shown to enhance high-intensity intermittent bouts of exercise and is safe to use within the recommended guidelines of about 5 g per day. Typically, athletes should not use creatine for extended periods of time, but rather cycle its use to periodically re-establish its sensitivity and efficacy. This can be done by cycling usage periods of about one to three months with periods without usage for one to three months. The use of legal stimulants before training or competition can also provide a performance-enhancing effect, as well as provide psychological enhancements. Individual tolerance to stimulants should be established before usage, and usage should follow a cyclic pattern as listed for creatine monohydrate. In addition, carbohydrate supplements may be effective during this time, as carbohydrate demands are relatively high and can be difficult to consume in large quantities. Carbohydrate supplements can be easily added to water or the athletes existing protein beverage and help the athlete maintain blood glucose levels during exercise, recover from training and competition, and help spare endogenous sources of carbohydrate during prolonged exercise.

It may also be worth noting that during the competition phase athletes commonly must travel to the competition site, which can be domestic or international. One common problem many athletes face is coming unprepared for competition and recovery nutrition, or often having to buy foods that are unfamiliar or foreign to them, such as a Brazilian soccer player having to buy food in China for the upcoming soccer match. Coaches and athletes should take the initiative of packing competition foods to help prevent many of the complications of travel. These foods should be familiar to the athlete so as not to cause gastrointestinal distress, but also be able to sufficiently fuel the athlete during pre-competition periods, competition periods, recovery periods, and down times. It is commonplace for many athletes to not pack any food items during travel, and then scramble to find something on the way, usually resulting in either suboptimal quantities of food, food choices, or some combination of both which of course can lead to poor performance and recovery. Coaches and sport scientists must take the initiative to remind athletes to pack food for travel, especially when traveling internationally, to avoid any unnecessary food related complications and minimize outside psychological stressors.

As athletes move chronologically throughout the training cycle into the competition phase, we see a shift from training for body composition and work capacity in the general preparatory phase, weight stabilization and increased sport specificity in the specific preparatory phase, and enhancing performance and recovery during the competition phase. We must remember that the training process is inherently stressful, and pairing a structured nutritional intervention is unfortunately no different. After a long season or intense training cycle, many athletes require some time off from structured training and nutrition which is commonly referred to as active rest. The last phase of the training cycle is the active rest phase and will help the athlete maintain a healthy relationship with the diet and training processes.

ACTIVE REST

The last major phase in the annual training plan is the transition or active rest phase, which is meant to link multiple training cycles together and completely alleviate the psychological and physical fatigue from intense competition and training cycles. Accordingly, dietary considerations must be factored into this fatigue as well, so the last phase of the annual nutrition plan is the active rest phase. Structured dieting can be a monotonous, stale, and oftentimes an unpleasant process. Just as an athlete needs time to unwind from the rigors of their sport, so too must they unwind from the stress of structured dieting. The purpose of this phase is to not only alleviate the aforementioned fatigue of the training cycle, but also to re-establish healthy and productive relationships with eating as well.

What does that mean in application? Do all the good diet habits go out the window, and athletes are allowed to just eat pizza and consume energy drinks? Of course not. Much of the rigidness or constrictive behavior can be eliminated in favor of eating quantities and types of foods that the athlete enjoys. However, this does not imply that all considerations go out the window. We still would like to encourage the athlete to still eat plenty of protein, have more carbs on days when they are more physically active, but after that give them the opportunity to satisfy food cravings and enjoy eating.

For the most part, coaches and athletes can eliminate any stressors relating to nutrient timing, food composition, and supplements. It would be great if the athlete was still able to make considerations for macronutrients and calorie balance, but not at the cost of dreading to weigh out portion sizes or having to do extensive math every time they eat. Additionally, the athlete should not be focused on maintaining a certain bodyweight or body composition. Especially for weight class sports, where bodyweight measurement and modification can be extensive, athletes should refrain from frequent bodyweight monitoring or obsession with bodyweight. Just as taking structured time off training can often lead to increases in desire to train, taking structured time off dieting can have a similar effect. Taking time off can allow the athlete to start the subsequent phases of dieting without any psychological hang-ups they developed during a previous cycle. One can think of the active rest phase as similar to that of having a weekend. No matter how much you like or dislike your job, just about everyone feels invigorated after a great weekend.

It is also important to remember that just as with training, the length of the active rest phase will depend on the level of the athlete and the severity of accumulated fatigue. Many sports will have inherently more difficult nutritional programming than others, such as mixed martial arts and virtually all combat sports, weightlifting, powerlifting, physique sports, gymnastics, and many others due to the need to compete in a weight class or optimizing power to bodyweight ratios. Other sports like rugby, basketball, tennis, and track and field are all uniquely challenging; however, they will not present the same degree of rigor and stress in the annual plan with regards to nutrition as weight class dependent sports. Accordingly, the amounts of time needed to alleviate that fatigue will likely vary from person to person and sport to sport.

Looking at the big picture of integrating training and nutrition we can see that the nutrition component largely plays a role in supplementing training to optimize conditions or taking advantage of the training conditions to make favorable changes in athlete's bodyweight or body composition. This is summarized in table 2.4.

Table 2.4 Goal summary for both training and nutrition throughout an annual plan

Training Phase	Preparatory		Competitive	Transition
	General	Specific		
Training Goals	Developing sport and training skills	Mastery of training and sport skills	Alleviating fatigue and elevating preparedness	Link multiple competitive cycles together
	Elevate work capacity and general fitness	Elevate sport-specific strength, explosiveness, and conditioning	Continued improvement or maintenance of fitness	Alleviate physical and psychological fatigue
	Basic sport tactics	Team and individual tactics	Perfecting skills and tactics	Maintain basic fitness
Training Phase	Body Composition Alteration	Stabilization	Recovery and Performance	Active Rest
Nutritional Goals	Gain muscle or lose fat	Achieve and maintain within 2% of competition bodyweight	Shift macronutrient priorities to carbohydrate	Reduce stress of dieting
	Develop basic diet and hydration routines	Refinement of eating and hydration patterns	Maintain euhydration	Maintain a reasonable bodyweight
	Consistent monitoring of bodyweight	Fatigue alleviation	Maintain lean body mass	Maintain lean body mass

PUTTING THE THEORY INTO PRACTICE

Now with a substantial foundation of theory and methodology behind our IP nutrition concept, we take a look at how to make it happen out there in the real world. Together we are going to work on developing microcycle, macrocycle, and annual nutritional strategies just as you would when making an annual training plan. In fact, the annual training plan is a great place to start! Normally when outlining a training program, it is preferable to start mapping out the competition periods first and then working our way back (or away in multistage plans) and our nutritional approach will be exactly the same—start at the end and work our way back to the beginning (or again away from competitions). However, for this first go around, we actually start backward with microcycle strategies and work our way up to macrocycle strategies. This will make it easier to understand for those with minimal experience in nutrition. Let's dive in!

MICROCYCLE STRATEGIES

Our microcycle strategies essentially deal with what the athlete eats day to day. This can obviously be different depending on how much activity differs day to day, and potentially on a macro and annual level, too, depending on if they need changes in body composition. First, we require a needs analysis to guide our nutritional and training strategies for this training cycle. Using normative standards as comparisons, the data collected through athlete testing and monitoring, and the insight from the coaches and athletes, we can determine not only the competition goals but ALL the strategies used from our IP concept as well.

Of particular importance in regard to nutrition are questions like:

1. How does the athlete compare to normative standards in bodyweight or body composition?
2. Is the athlete at or near their ideal competition weight?
3. Does the athlete need to gain muscle/maintain muscle while reducing fat?
4. How do these three questions affect not only this training cycle, but also subsequent training cycles in the long term?

It is an easy mistake to get caught up in the short-term and focus on maximizing performance right here and now, getting athletes peaked for the upcoming season. However, IP reminds us that the long-term is just as important as the short-term, and training plans should address strategies for this season, two years from now, five years from now, and potentially beyond. For example, you might have a pretty good freshman linebacker weighing 93 kg (205 lb) having a great camp and pre-season, but what if you could train him up to 111 kg (245lb) by the time he was a senior? That could be one of the differences between becoming an NFL caliber athlete and not. Hence, our needs analysis should be telling us what we need to do now, but also giving a friendly reminder of where we are going in the future.

ESTIMATING DAILY CALORIE NEEDS

Once we have determined the general direction of our athlete from our IP approach, we can start working on putting together a microcycle strategy. The first step is often the most overlooked step, yet is simultaneously the most powerful tool in our nutrition toolkit: Determining the athlete's state of calorie balance, or finding his or her eucaloric state. This is truly the meat and potatoes of our nutrition plan and, very much like how having an annual plan, allows for our periodization concept to become more easily organized, so too does determining calorie balance on setting up microcycle strategies. Once we have determined maintenance level calories we can:

1. Potentially adjust those calories to meet the needs of body composition alteration
2. Portion the daily calories into the requisite amount of daily proteins, carbs, and fats
3. Split the meals based on our nutrient timing principles
4. Determine the macronutrient quantities of each meal individually
5. Determine the actual food content of each meal individually

This may seem straightforward, but this step is often overlooked because it is not a trendy or popular point of discussion and can be confusing and frustrating to figure out. How do we determine daily calorie needs? There are several methods we can use, all with varying degrees of accuracy and practicality. One of the most common and simple way is using daily calorie estimating equations. There are a number of these available online, and generally they are a great starting point for those who are not sure where to begin. One note is the Harris-Benedict equation, which estimates daily calories through an estimation of the basal metabolic rate from an athlete's gender, age, height, and weight and then factors in a daily energy expenditure estimate based on relative activity levels. It is certainly not perfect and represents an estimate at best, but again an excellent place to begin for those delving into this for the first time. Similarly, there are numerous publications in journals, textbooks, and other sources outlining energy expenditure of different activities and varying intensities as a work rate usually expressed in kilocalories per minute or per hour. These can be found in exercise physiology, sport science, and sport nutrition textbooks and function as a great supplemental tool to estimate calorie needs.

Moving into more accurate but maybe less practical methods of calorie estimations are the use of different forms of calorimetry. Indirect calorimetry methods such as gas exchange are probably the most common method of calorimetry used for this purpose and can be found in almost any exercise physiology lab in the world and is becoming increasingly prevalent in modern medicine as well. Gas exchange is a method of estimating substrate utilization, and therefore calories, through the measurement of inhaled oxygen and exhaled carbon dioxide. For exercise, it is typically performed at a steady state pace on a treadmill or cycle ergometer or can be modified with a stepwise or ramp protocol for a graded maximal exercise test for variables such as $VO2_{max}$ and maximal heart rate. It can also be done at rest to give a measurement of resting metabolic rate. This method is certainly more accurate than using estimated equations; however, it does come at the cost of requiring access to lab space and potential monetary expense. Direct calorimetry methods can be used; however, these require the use of very large temperature chambers which measure direct changes in heat. Currently, these methods are following the unfortunate trend to become very expensive storage closets, and for sport nutrition purposes they cannot really mimic the actions of most sports due to enclosed nature.

Arguably the most practical and cost-effective method of estimating daily calories is something that does not require any labs or fancy equations, and in fact due to advances in technology and fitness trends has become easier than ever before. It is simply recording daily intake of food and monitoring trends in bodyweight. This must be done diligently and honestly, either manually by recording everything that was eaten day to day or logged in an application which can automatically calculate daily calories and macronutrients. See figure 2.6 for an example of what a daily food log might look like. Calories should be logged daily for one to two weeks so that a seven-day weekly average can be calculated. This number will represent on average how many calories the athlete is eating per day, and if carried out over several weeks should also include normal day-to-day variability.

Name	Servings	Calories	Carbs	Fiber	Fats	Protein	Time
GV Raisin Bran Crunch	2 cups	380	88	8	2	6	Breakfast
Skim milk	2 cups	182	24	0	2	16	Breakfast
Gatorade	2 scoops	450	126	0	2	0	Workout
ON GSW	2 scoops	260	8	0	3	48	Workout
Oikos fruit Greek yogurt	1 container	130	21	0	0	12	Post-workout
Brown rice (cooked)	1 cup	216	45	4	2	5	Post-workout
Chicken breast	8 oz	220	0	0	8	40	Post-workout
Birds Eye California Blend	1 cup	25	4	1	0	0	Post-workout
GV whole wheat bread	4 slices	280	52	8	2	12	Lunch
Oikos fruit Greek yogurt	1 container	130	21	0	0	12	Lunch
GV whole wheat bread	4 slices	280	52	8	2	12	Dinner
Gatorade	2 scoops	450	126	0	0	0	
ON GSW	2 scoops	260	8	0	3	48	
GV Raisin Bran Crunch	2 cups	380	88	8	2	6	Snack
Skim Milk	2 cups	182	24	0	2	16	Snack

Figure 2.6 Sample foods logs done by a fitness app or using a spreadsheet

Bodyweight measurements have a large degree of day-to-day variability for a number of different reasons, as seen in figure 2.6. Hence, it is important not to put too much stock into any one single measurement, but rather methods like trend analysis and statistical process control will give a much better idea of what is actually happening. We recommend taking a baseline measurement of bodyweight two to three times per week, measured consistently under nearly identical conditions for at least two weeks, and cross-referencing this trend with a seven-day weekly average of calories consumed per day. If the trend in bodyweight is generally flat, then the calories per day are probably on point with a eucaloric state. If the trend is upwards or downwards, then by definition they are in a hyper- or hypocaloric state, respectively. If we are seeking to find the eucaloric state, one can simply modify the calories in small chunks at a time (± 250-500 kcal) week to week until bodyweight stabilizes. This calorie measurement then becomes our baseline daily calories which all other macronutrients are based on. This, of course, is subject to change depending on activity levels, bodyweight changes, and deliberate body composition alterations.

DETERMINING MACRONUTRIENT VALUES

The hard part is over. We have determined out how many calories our athlete needs per day to meet his or her goals for this training cycle. The next step will be divvying up those calories into the appropriate amount of each macronutrient per day. Now you may have seen various pie charts or relative scales stating that athletes should eat 25% protein, 65% carbs, and so on. However, such scales are far too generalized and do not account for individual differences or daily activity levels. Instead, we look back at the theory we discussed on macronutrients earlier to figure out how much of each we need.

When talking about body composition alteration and performance protein, we must determine the number of macronutrients we must account for first. Dietary protein literally makes up the building blocks of skeletal muscle tissue and is essential for not only building muscle, but also repairing damaged tissues from intense voluminous training sessions. The amount of protein needed for each athlete is largely dependent on their LBM content and activity. Pure strength/power athletes who generally carry more LBM tend to have slightly higher protein recommendations; field sport athletes tend to be right in line with our general previous recommendations; and endurance athletes tend to only require a moderate protein intake with a substantially higher carbohydrate requirement. Note: These are generalities, and individualization should always be taken into account. Is it possible to have an exceptionally muscular female soccer player or a track sprinter who is not as muscular as his peers? Absolutely, and that is a reason why activity alone may not be the best indicator of protein requirement; rather the athlete's LBM or bodyweight may provide better insights.

We start with a simple example, then we can explore more options later once we have the basics down. For instance, we have a male strength athlete weighing about 100 kg who is taking in about 3650 kcal per day. In our previous discussion, we recommended a protein intake of about 1.8 – 2.2 g per kg (0.8 – 1.0 g per lb) of bodyweight per day. For our 100 kg (220 lb) strength athlete that would equate to about 220 g of protein per day. Protein has a caloric value of about 4 kcal per gram, so this 220 g of protein would account for about 880 kcal. Make a mental note because we will come back to those numbers in a little bit. The next item on our macronutrient priority list will be carbohydrates.

In our previous discussion on carbs, it was determined that carbs should mainly be prescribed based on daily activity levels. Lower activity levels will result in lower carbohydrate intake and vice versa. Our strength athlete trains 1-2 times per day for an accumulated 90 minutes. Based on our previous recommendations that would result in a carbohydrate intake of about 2.2 – 3.3 g per kg (1.0-1.5 g per lb) of bodyweight per day. To make things easy, let us use 3.3 g per kg per day (~1.5 g per lb). This would result in a daily carbohydrate intake of about 330 g, and just like protein carbohydrate has a caloric value of about 4 kcal per gram, accounting for about 1320 kcal of our daily total.

Up to this point we have accounted for calories, protein, and carbohydrates. Last on our macronutrient list will be fats. In our previous discussion, it was determined that fats did not have any real direct relationship with workloads or anything outside of serving as an excellent calorie buffer after protein and carbohydrate demands have been met. To determine how many grams of fat are needed per day, we will simply take the calories of protein and carbs and subtract them from our daily calorie needs. For our strength athlete requiring 3650 kcal per day, we will subtract calories from protein (880 kcal) and calories from carbohydrates (1320 kcal) to find the remaining calories per day (1450 kcal). Fat has a caloric value of about 9 kcal per gram, so we can divide those remaining calories by 9 to get the grams of fat needed per day (~161 g).

At this point, we have accounted for calories and macronutrients, which are by far the two most powerful considerations we can manipulate. With the information we have gathered thus far, we can throw all other considerations by the wayside and still be roughly 80% on point with our diet, which is still very powerful. At this point, we could easily start using flexible dieting techniques such as If It Fits Your Macros (IIFYM), or we could just split the macronutrients evenly throughout the day for our desired number of meals (4-7). However, this could potentially (but not necessarily) leave out that last 20% or so of success from factors such as nutrient timing, food composition, and supplements. We must keep pushing a little bit further and delve into nutrient timing.

In our previous discussion of nutrient timing, it was determined that protein seems to be best split it into multiple smaller meals throughout the day, carbohydrate intake was denser around training times, and fats were the polar opposite to carbohydrate intake with a low density around training times and higher throughout the day. With our example strength athlete, we start by portioning out the required dietary protein. His protein prescription per day was 220 g. There does not appear to be any real benefit to portioning out protein differently throughout the day, so we can simply take the total number of grams per day and divide by the desired number of meals per day, and for this example, we use six meals per day. This would give us a per meal dose of about 37 g of protein, which we would spread fairly evenly throughout the day separated by about three to five hours.

Carbohydrate, on the other hand, will be emphasized around training times and, of particular importance, the post-exercise period. Generally, a good place to start is the intra-workout and immediate post-workout meals. For most normal training sessions lasting longer than 60 minutes, it's generally recommended that athletes consume some intra-workout carbohydrate sources, usually high glycemic fluid sources such as sports drinks and fruit juice, to help maintain blood glucose levels and exercise intensity, as well as dampen the effects of fatigue. For most competitive athletes, this is fairly commonplace outside of light workouts and recovery sessions, so this is an easy place to start. How much carbohydrate? An excellent place to start is using the previously stated glycogen repleting protocol of about 1 g of carbohydrate per kg of bodyweight per hour. For our 100kg strength athlete, this would be about 100 g of carbohydrate during the workout which can be consumed continuously or intermittently. That same value can be used once again for the immediate post-training meal, which should occur as quickly as possible following the training session, preferably no more than an hour after completion.

In terms of pre-exercise carbohydrate, there is no right or wrong answer here as this is highly dependent on how proximal the meal is to the exercise session and the individual's tolerance to pre-exercise meals. It is generally recommended that pre-exercise meals should occur about one to three hours prior to the session, and the size of the meal should be directly proportional to how far away it is—in other words, the closer to training, the smaller the meal, and vice versa. A very simple starting place will be to take the glycogen repletion protocol value and divide by 2. For our 100kg athlete, that would be equal to 50 g of carbohydrate pre-exercise, 100 g during exercise, and 100 g immediately following the session. The remaining carbohydrate for the day can be tapered down in the post-exercise period. Fats will follow an almost identically opposite trend as carbohydrate. Table 2.5 shows what this might look like in the real world.

Table 2.5 Nutritional examples for one day training for a 100 kg male strength athlete

Training in the Morning				Training in the Afternoon				Training in the Evening			
Meal	Protein	Carbs	Fats	Meal	Protein	Carbs	Fats	Meal	Protein	Carbs	Fats
Pre	37 g	50 g	10 g	Breakfast	37 g	0 g	55 g	Breakfast	37 g	0 g	55 g
Intra	37 g	100 g	0 g	Pre	37 g	50 g	10 g	Lunch	37 g	30 g	55 g
Post	37 g	100 g	10 g	Intra	37 g	100 g	0 g	Pre	37 g	50 g	10 g
Afternoon	37 g	50 g	31 g	Post	37 g	100 g	10 g	Intra	37 g	100 g	0 g
Evening	37 g	30 g	55 g	Evening	37 g	50 g	31 g	Post	37 g	100 g	10 g
Bedtime	37 g	0 g	55 g	Bedtime	37 g	30 g	55 g	Bedtime	37 g	50 g	31 g

In table 2.5, we can see some different options using the same numbers previously discussed for our 100kg male strength athlete. The column on the right shows a sample daily plan when training in the morning, the middle shows training in the afternoon, and the right shows training in the evening. Notice that protein was straightforward, and after mapping out the pre-, intra-, and post-carbohydrate meals, the rest falls in line very nicely. Now keep in mind this is a fairly simplified version; however, the process to go about this and the resulting trends are exactly the same.

In reality, at this point, the hard part is over. The basics and most powerful factors have been covered and all that remains are the finer points of which foods and potentially supplements to choose. Looking back at our food composition section we can start choosing proteins, carbs, and fats to suit the needs of each meal. Although initially this may seem overwhelming and complicated, after time for familiarization and practice this quickly becomes second nature. For some, understanding how to read food labels properly can be a challenge, while for others it can be challenging to find information on items that do not come with a standard label such as fruits, vegetables, certain meat products, meals at restaurants, and so on. Luckily there are a vast number of websites, such as nutritiondata.self.com, and many others which have consolidated huge databases of food sources and their respective serving sizes, making portioning much easier than previously.

As previously discussed, the actual types of foods eaten really are not of critical importance but remain a worthwhile consideration when trying to maximize the effects of a nutrition program. In the previous example, all meals that are intra-workout or immediate post-workout meals will probably be rich in lean meats or dairy, whole grains and whole wheats, fruits, vegetables, nuts and natural nut butters, olive oil, and similar foods. Intra-workout meals will generally be in liquid form and consist of fast digesting food sources such as high glycemic carbohydrates found in sports drinks and juices and protein powders like whey and its variations. Post-exercise meals can often be a blend of both solid and liquid, and fast and slow digesting sources. Particularly when the dosage of carbohydrate is large, it can be difficult to consume exclusively complex sources, so high-to-moderate GI sources like cereal, fat-free gummy snacks, white bread with jams and jellies, and others can elicit the desired effects while still being able to be consumed in large quantities without excessive discomfort. Protein sources can also be solid or liquid, with dairy products like chocolate milk or Greek yogurt being excellent choices at these times.

We can now use this same systematic approach to evaluate athletes under different circumstances. Table 2.6 lists some examples of different athletes and their respective changes in dietary needs. Notice each athlete has a different body size affecting how much protein they generally need per day, and each has a unique training plan and differing total training volumes affecting how much carbohydrate they need per day. It is important to reiterate that the volume of exercise is the primary determinant of how much carbohydrate is consumed per day. One of the mistakes that people often make in this assessment is confusing volume and intensity. For instance, Christian a weightlifter works out at very high intensities, meaning the power output and loads lifted are generally very high, whereas Michelle, the triathlete, generally trains at much lower intensities but ultimately completes a much higher amount of work per session. That is not to say that Christian's workouts are not hard or that Michelle's are inherently easier, but rather the total amount of work performed was significantly different between the two. Thus, the relative difficulty of the exercise is secondary to the total amount of exercise performed when considering carbohydrate.

Table 2.6 Examples of daily nutritional needs by sport

Athlete	Melissa	Christian	Andy	Michelle
Sport	Mixed Martial Arts	Weight lifting	Rugby	Triathlon
Gender	Female	Male	Male	Female
Bodyweight	61 kg	87 kg	110 kg	57 kg
Training sessions per day	2–4	1	2	1–3
Training times	AM/PM Fight, Lift noon	PM Lift	AM Lift, Afternoon Rugby	Varies by day
Total training time per day	2–4 hours	1–2 hours	2.5 hours	3+ hours
Estimated daily calories	3,000 kcal	3,020 kcal	4,035 kcal	2,655 kcal
Daily protein	1.8 g per kg = **110 g**	2.2 g per kg = **191 g**	2.2 g per kg = **242 g**	1.5 g per kg = **86 g**
Daily carbohydrate	5 g per kg = **305 g**	2.2 g per kg = **191 g**	4.4 g per kg = **484 g**	7.5 g per kg = **428 g**
Daily fat	$(3{,}000-1{,}660)\cdot 9^{-1}$ = **149 g**	$(3{,}020-1{,}528)\cdot 9^{-1}$ = **166 g**	$(4{,}035-3{,}064)\cdot 9^{-1}$ = **108 g**	$(2{,}655-2{,}056)\cdot 9^{-1}$ = **67 g**
Meals per day	7	5	6	6

MACROCYCLE STRATEGIES

Now that we have outlined the basic day-to-day needs in terms of all our major dietary factors, we can begin to formulate a systematic macrocycle strategy approach leading up to our competitive phases. Although our microcycle strategies revolved around fairly simple concepts like energy expenditure, body size, and activity levels, our macrocycle strategies will be heavily rooted in our needs analysis, the testing and monitoring processes we have established for our athlete, and the competition schedule. In order to understand how to integrate our annual plan with nutritional considerations, we also have to have a strong grasp on where our athlete currently stands relative to where they need to be. This is true not only for things like body size and composition, but also performance characteristics like strength and power as well.

Just like setting up an annual training plan, this process can be streamlined or arduous depending on how well you organize. We are going to follow the same general recommendations of writing an annual training plan by starting at the major competitions and working our way backward to the start of the training cycle and forward to any active rest or transitional phases. The first step will be to identify the major competitions or competitive periods and outline the recovery and performance phase in the annual plan. Virtually, this should always be in alignment with the competitive phase of the training plan.

At this point, we are assuming that the athlete will be at their competition bodyweight and body composition, outside of some minor changes that might be used for weigh-in purposes. Next, we must outline when the bodyweight will initially become stabilized, which should be about one to three months out from the competitive phase. The stabilization phase is intended to hold the athlete within ± 1-2% of their competition bodyweight and should closely follow the specific preparatory phase of the training plan. The length of the stabilization phase depends on how far away the athlete is from their competition bodyweight, so if the athlete does not have a substantial amount of weight to gain or lose, then the stabilization phase can be upwards of two to three months. In contrast, if the athlete needs more time to gain or lose weight, then the stabilization phase will be relatively short lasting as low as one to two months. Skipping this step is a major "beginner" mistake for a lot of coaches and athletes who push bodyweight and body composition changes all the way into competitions. A smart coach, athlete, or trainer will address these changes early on and reap the benefits of more productive training and recovery leading into the competitive phases.

Last but certainly not least in this area of the nutrition plan is the body composition alteration phase, which should closely mimic the general preparatory phase of the training plan. This phase should generally be the furthest away from major competitions due to its inherently more stressful nature. The length of this phase can vary depending on how much weight an athlete is planning on gaining or losing. Generally, it is recommended to not exceed 12 weeks of continuous hard body composition alteration at a time, and athletes who require more than this should break up this phase with periodic stabilization phases to alleviate dietary fatigue and allow their homeostatic set points to adjust. There are instances where some athletes may not be striving to make any body composition changes, and in that case, they can substitute a stabilization phase during this portion. Generally speaking, most athletes, regardless of sport, have room for improvement in the realms of gaining muscle or losing fat. If the athlete reaches their targeted bodyweight earlier than expected, then they can move into stabilization. However, coaches and athletes should be cognizant of the ill effects of gaining and losing weight faster than 1-2% in bodyweight per week.

In the direction of competition, we have the active rest phase, which is used as either a recovery period or as a link to multiple seasons or different annual plans. This phase will generally come after the major competitive periods of the training cycle and should virtually follow the active rest and transition portions of the training plan. The length of this phase depends on the amount of allostatic load (see chapter 8) the athlete is carrying, but generally lasts for about one to four weeks. Typically, this will be followed by a body composition alteration phase if the athlete is moving into an off-season or non-competitive portion of their training cycle or move directly into stabilization if competitions are reappearing in the next several months.

ANNUAL PLANNING STRATEGIES

The annual strategies often revolve around two central concepts, which the authors believe to be an inherent part of Integrated Periodization:

1. Continued growth and athletic development
2. Allostatic management

Oftentimes coaches and athletes get caught up in the here and now, thinking how they can just do their best in the upcoming season. This is not an inherent flaw as continued athletic success also includes the present, but often this thinking comes at the expense of long-term athletic development. For highly competitive athletes, one of the major limitations to nutritional periodization is simply time. For sports with long seasons or multiple sub-seasons within the year like hockey, track and field, rugby, and others, there simply might not be a lot of time allocated for things like body composition changes or complete recovery during the training cycle. For some sports that might not be a major problem, but for others where normative standards for size and strength continue to rise with the level of play, this can potentially be a differentiating factor for success. This is especially true for sports such as high school football leading into Division 1 football, thereby leading into the National Football League.

When looking at long-term strategies for training such as a quadrennial plan, it is not uncommon for coaches and athletes to place slightly different emphases at certain times throughout the athlete's development. For example, a coach may choose to spend more time on hypertrophy throughout the year with younger or less developed athletes, and more time on speed and power development with more trained or mature athletes. Some strategies that result in not fully peaking, continuing overload training through competitions, emphasizing hypertrophy and basic strength characteristics may actually cause short-term decreases in performance, with the hopes of long-term gain several years later down the road. Specific strategies such as peaking and tapering, reducing training volumes, emphasizing power and speed characteristics may have a very positive effect on performance and fatigue management in the short-term, but may not be providing as much development in the long-term.

What this implies is that coaches should take the time to think about the short-term, intermediate-term, and long-term strategies that are placed on athletes. Accordingly, the issue becomes how to best adjust the nutrition plan to meet training and competitive goals. This is a good time to remember that sport nutrition is meant to supplement training programs, and training goals need to be well established and outlined before integration of nutrition is really possible. Imagine asking your athletes to begin eating a surplus of 500-1000 calories per day, but the only training prescribed was sport practice and sprints. Will they gain weight? Yes. Will a large portion of that weight be muscle mass? Probably not, because the training and nutrition goals were simply not synchronized. Training and nutritional strategies must be integrated into the short-, intermediate-, and long-term.

The following questions can start to help formulate annual strategies of Integrated Periodization:

1. How close is my athlete to reaching normative standards of size, strength, or leanness of their peers?
2. How much time has been allocated to recovery, both physical and psychological?
3. How close is my athlete to reaching normative standards of size, strength, or leanness within their future competition levels (collegiate, professional, Olympic, etc.)?
4. What is the goal of the current training cycle?
5. What are the goals for the next 2-5 training cycles?
6. How much time can be allocated for body composition alteration, stabilization, and active rest throughout the annual plan?

SUMMARY

Sport nutrition is an often underused or undervalued component of the training process. Sport nutrition is distinct from health or clinical nutrition in that it solely seeks to enhance sport performance and does so in the areas of providing sufficient energy, enhancing body composition, and promoting the recovery–adaptive processes. Sport nutrition should not be treated as a separate outside component of the training process, but rather integrated with physical and psychological training that make up the sport.

Not all components of nutrition are equally effective and powerful. Managing energy balance and calories consumed is by far the single most powerful variable for body composition and performance enhancement. Contrary to what many trends in fitness would have you believe, what you eat is not nearly as significant as simply how much you eat relative to your training load. Within calories, the relative contribution of each macronutrient plays another large role. The amount of protein needed per day is generally based on athlete lean body mass, the amount of carbohydrate is generally based on net activity levels, and the amounts of dietary fat is generally used to buffer calorie demands after protein and carbohydrate needs are met. Nutrient timing, although complex and intriguing to many, plays an important role but is not nearly as powerful as calories and macronutrients. Food composition considerations and dietary supplements are fairly weak in terms of their effect on performance; however, these can be used for fine tuning sport nutrition practices.

Periodization of nutrition largely mimics periodization of training, and the phases involved in both training and nutrition are inherently linked in our Integrated Periodization concept. The general preparatory phase of training is linked to the body composition phase of nutrition where athletes are striving to gain muscle or lose fat. The specific preparatory phase of training is linked to the stabilization phase, where athletes achieve a competition bodyweight and maintain it within ±1-2% of their ideal competition bodyweight. The competitive phase is linked to the recovery and performance phase, which puts a larger emphasis on carbohydrate consumption to ensure optimal performance and recovery. The active rest phase is used for both training and nutrition to provide athletes with a complete physical and psychological recovery from hard dieting, training, and competition.

Chapter 3

PSYCHOLOGICAL PREPARATION AS AN INTEGRAL PART OF ATHLETIC TRAINING

Boris Blumenstein, PhD / Iris Orbach, PhD

A strong mind may not win an Olympic medal, but a weak mind will lose one.
–Si, Statler, & Samulski, 2014, p. 503

The awareness to the importance of psychological preparation (PP) among coaches and athletes has grown tremendously in the last three decades. Scientifically, based on an analysis of research results that investigated the effectiveness of PP in competitive individual and team sports, a positive performance effect was found in more than 85% of the studies (e.g., Gould & Carson, 2007; Greenspan & Feltz, 1989; Meyers, Whelan, & Murphy, 1996; Vealey, 1994, 2007; Weinberg & Comar, 1994; Weinberg & Gould, 2015).

From an applied perspective, the majority of athletes indicated the importance of PP; however only 44% made frequent use of psychological strategies and techniques (Heishman & Bunker, 1989). In addition, the use of psychological techniques by athletes occurred mostly in competitions compared to practice settings, which does not allow for optimal PP (Frey, Laguna, & Ravizza, 2003). The full potential of PP can be seen when it is part of the athlete's overall preparation—physical, technical, tactical, and psychological (Balague, 2000; Blumenstein et al., 2007; Bompa, 1999). Unfortunately, coaches and athletes are not aware of the potential of systematic and integrated psychological preparation within the daily training program and preparation phases.

The chapter's four purposes are these: first, to introduce the fundamentals of psychological training; second and third to discuss the psychological skills and the strategies associated with peak performance; and finally, to establish an original scientific and practitioner approach which integrates the psychological preparation with the periodization principle.

FUNDAMENTALS OF PSYCHOLOGICAL PREPARATION

Psychological preparation has become a major foundation of any sport program that aims to teach the athletes psychological techniques/strategies and prepare them for the competition event (Singer & Anshel, 2006; Weinberg & Williams, 2015). Psychological preparation is provided by a sport psychologist/consultant who can take one of two different philosophical approaches: the educational approach and the clinical approach. The educational approach is oriented from systematic psychological skill training, necessary for the athlete's peak performance in sport and personal well-being. The objective of the clinical approach is to provide therapeutic assistance for any dysfunctional personality processes and behaviors of the athlete (Vealey, 2007). Scientific and applied evidence has indicated that the majority of athletes working with sport psychologist benefit more from the educational approach as opposed to a clinical approach (Weinberg & Williams, 2015).

Effective psychological preparation cannot stand alone and should be integrated with all other athletic preparations (see chapter 1). Psychological preparation strengthens the physical, technical, and tactical preparation in each training session and practice. Clear relationships currently exist only among the physical, technical, and tactical preparations, while the link to the psychological preparation is somewhat neglected (Blumenstein et al., 2005; Blumenstein et al., 2007; Lidor, Blumenstein, & Tenenbaum, 2007). In order to achieve the athlete's peak performance, all four components of preparation should be interrelated (see figure 3.1).

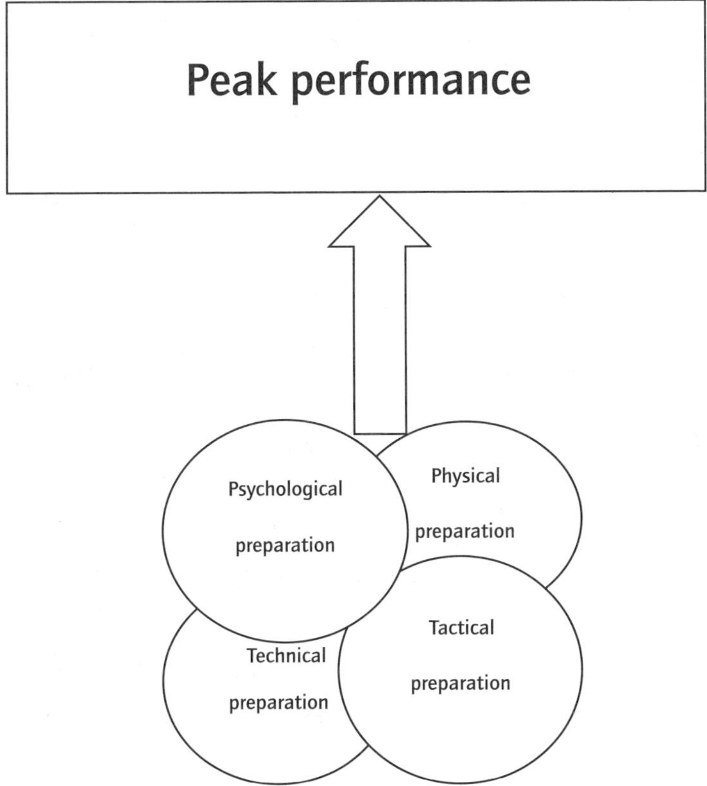

Figure 3.1 Desired relationship among the athletic training preparations

A major aim of psychological preparation is to teach and practice psychological skills within a psychological skill training (PST) program that can be applied in practice sessions and competition. PST has been defined as the systematic and consistent practice of psychological skills for the purpose of athletic peak performance. A successful PST program is based on the following fundamental principles.

PRINCIPLES OF PSYCHOLOGICAL SKILL TRAINING

PRINCIPLE 1: SYSTEMATIC PSYCHOLOGICAL SKILL TRAINING

Psychological skills are learned and practiced systematically, just like physical skills (Blumenstein et al., 2005; Weinberg & Williams, 2015). PST can be provided in different manners based on a different time frame: short-term intervention is provided for a specific situation such as a concrete competition; periodical intervention is based on training with voluntary pauses while working with the athletes during a specific time frame such as a competition phase; systematic long-term psychological training is provided in a similar way to other preparations throughout the athlete's training season. An optimal positive impact can be seen when adopting a long-term intervention, which may include 1–3 weekly sessions, depending on the athlete's skills proficiency and his/her needs.

PRINCIPLE 2: PLANNING PSYCHOLOGICAL SKILL TRAINING

PST should be planned according to the periodization principle of an athlete's training. More specifically, psychological skills should be adjusted to the training phase, while the main objective is using the skills in practice and competition. Designing PST interventions provides assurance for a consistent and powerful intervention (Balague, 2000; Blumenstein & Orbach, 2012a,b; Holliday et al., 2008).

PRINCIPLE 3: INTEGRATION OF PSYCHOLOGICAL SKILL TRAINING

PST should be integrated with all other elements of athletic preparation—the physical, technical, and tactical (Balague, 2000; Blumenstein et al., 2005; Blumenstein & Orbach, 2012a,b). The integration of PST within the different preparations is likely to result in high-level sport achievements. This process allows the transfer of psychological skills from laboratory to field and ultimately puts it all together—physically and mentally—in the competitive setting.

PRINCIPLE 4: COLLABORATION WITH THE COACH AND MEDICAL STAFF

Cooperation between the sport psychology consultant and the coach/medical staff is critical for a successful integration of PST within the training process. Therefore, sport psychology consultants should be part of the professional staff, understand the fundamentals of sport training, and provide educational psychological materials to the coach and staff (Blumenstein & Orbach, 2012a; Gould & Carson, 2007). Unfortunately, PST is often neglected by the coach due to a lack of knowledge, perceived lack of time, or a belief that psychological skills are innate and cannot be taught (Weinberg & Gould, 2015).

PRINCIPLE 5: LINK BETWEEN SCIENTIFIC RESEARCH AND TRAINING PRACTICE

Successful PST should be developed and based on norms and hard data, which can be defined by scientific research. Scientific data can serve as a theoretical basis and provide guidelines for psychological application during the training process, formulate future perspectives, and provide clarity and confidence in the PST program. This should lead to a systematic construction of the program and a powerful PST concept (Blumenstein & Orbach, 2012a; Holliday et al., 2008; Kellman & Beckmann, 2003; Vealey, 2007).

PRINCIPLE 6: ETHICS AND MORALS IN PST APPLICATION

Ethics and morals for sport psychology consultants are aimed toward providing a framework for those who conduct research and perform applied work in the field of sport psychology. Major ethical codes within the field of sport psychology are competence, confidentiality, integrity, responsibility, and respect (Hackfort & Tenenbaum, 2014; Oliver, 2010). This code of ethics, tailored specifically for sport psychology, is a "vital aspect of the overall professionalization of the field" (Ziegler, 1987, p. 138).

PSYCHOLOGICAL SKILLS ASSOCIATED WITH PEAK PERFORMANCE

During the past three decades, sport psychology researchers and applied sport psychologists have been focused on psychological skills and strategies that are essential for enhancing athlete performance (Gould, 2002; Gould & Damarjian, 1998; Gould & Maynard, 2009; Henschen, 2005; Orlick & Partington, 1988; Vealey, 1988, 2007). For example, Vealey (2007) has described four multiple types of important mental skills for the athlete's and coach's success, including foundation skills, performance skills, personal development skills, and team skills. Most research in sport psychology has involved mainly four basic psychological skills–which have beneficial effects on performance–imagery, goal setting, relaxation/activation, and self-talk (Eccles & Riley, 2014).

It should be emphasized that each sport has unique demands for the psychological skills that may guarantee peak performance in this sport. Taking these studies together, the following common psychological skills and strategies are typically used in different sports to achieve peak performance:

- Relaxation and self-regulation of arousal (Statler & Henschen, 2009; Vealey, 2007; Weinberg & Gould, 2015)
- Concentration/attentional focus (Moran, 2003, 2010; Vealey, 2007; Vernacchia, 2003)

- Self-confidence (Feltz & Oncu, 2014; Gould, 2002; Vealey & Vernau, 2010)
- Self-talk (Hardy & Zourbanos, 2016; Hatzigeorgiadis, Zourbanos, Latinjak, & Theordorakis, 2014; Vealey, 2007)
- Imagery (Cumming & Williams, 2014; Suinn, 1993; Vealey, 1988, 2007; Vealey, & Forlenza, 2015; Vealey & Greenleaf, 2006)
- Goal-setting (Gould, 2015; Vealey, 2007; Weinberg & Butt, 2014)
- Biofeedback training (Beauchamp, Harvey, & Beauchamp, 2012; Blumenstein & Orbach, 2014).

To achieve peak performance athletes should practice the learned mental skills by using psychological strategies within PST (see Figure 3.2).

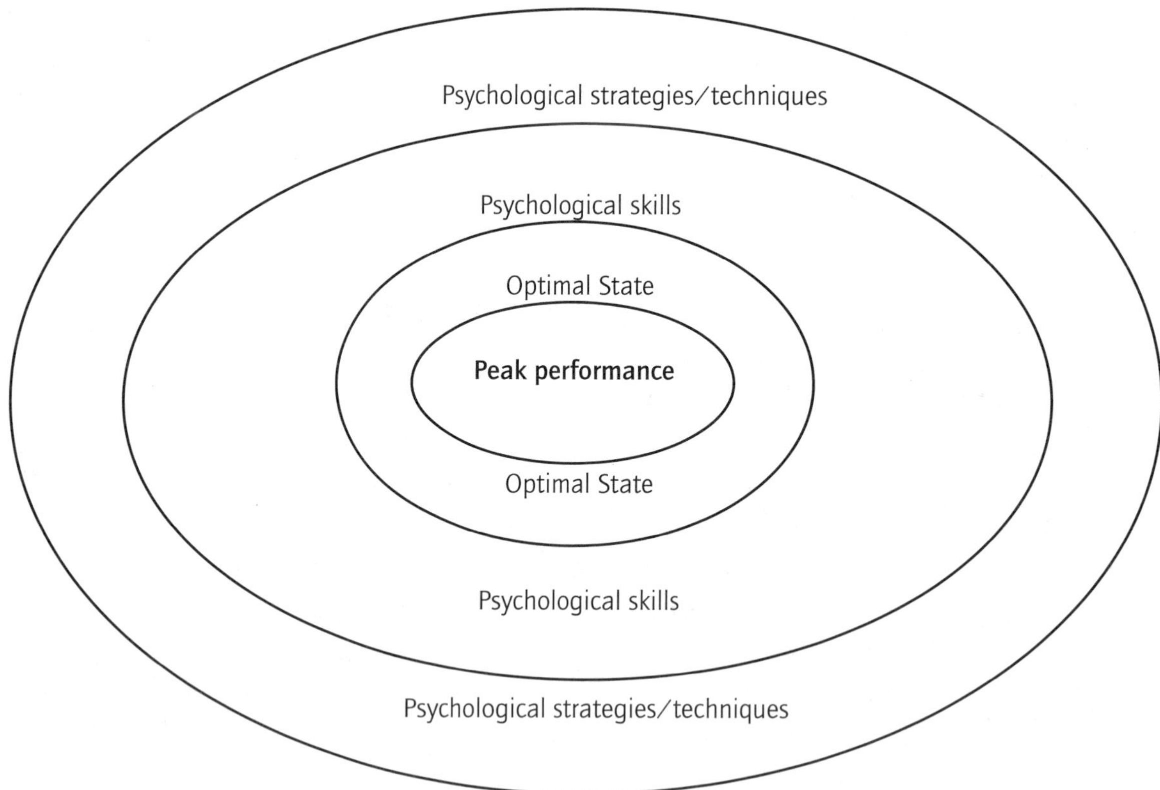

Figure 3.2 Schematic description of psychological components of peak performance

PSYCHOLOGICAL STRATEGIES ASSOCIATED WITH PEAK PERFORMANCE

PST can include a number of popular and well-known psychological strategies/techniques, such as relaxation, imagery, goal setting, self-talk, biofeedback training, concentration, and a precompetition routine (see Acharya, & Moris, 2014; Blumenstein & Orbach, 2014; Cotterill, 2010; Henschen, 2005; Lidor, 2007; Moran, 2010; Vealey, 2007; Vealey, & Forlenza, 2015; Weinberg, 2010). These techniques are those most widely used by sport psychology consultants, and they have shown significant positive effects on athletic performance. Each of these techniques can be used by itself or as part of a psychological intervention package (Blumenstein & Orbach, 2012a,b; Vealey, 2007). Moreover, these psychological strategies are characterized by individual content and context, which is determined by the different sport demands and sport situation. All of the above should be linked and coordinated with the training phases. For example, the concentration style in shooting is different than the concentration style in basketball; relaxation in the preparation phase and relaxation in the competition phase have different goals and therefore should be practiced accordingly. Following are descriptions of the major psychological strategies/techniques that should be learned and practiced within PST.

Relaxation is a popular technique in sport associated with the ability to regulate physical energy and the level of arousal during highly pressurized situations. It has been reported that relaxation may facilitate recovery from exercise (Acharya & Morris, 2014; Blumenstein & Orbach, 2012b; Gould & Udry, 1994; Vealey, 2007). The most popular and well-known relaxation techniques are "muscle-to-mind" techniques (physical relaxation strategies), such as breathing exercises and progressive muscular relaxation, introduced by Jacobson (1938), and "mind-to-muscle" techniques (mental relaxation strategies), such as autogenic training (Schultz, 1932), meditation, and yoga. The effectiveness of these relaxation techniques can be found in numerous studies within the field of sport psychology (e.g., Vealey, 2007; Weinberg & Gould, 2015). The objectives of the relaxation techniques are to reduce anxiety symptoms, regulate arousal levels, and benefit recovery purposes. Relaxation can be used along with other psychological strategies, such as self-talk and imagery, as part of a multimodal intervention package in PST (Vealey, 2007).

Imagery involves internally experiencing a situation that mimics a real experience, without experiencing the real thing. It is similar to a real sensory experience utilizing all of the senses (e.g., seeing, feeling, or hearing), but the entire experience occurs in the mind (see Cumming & Williams, 2014; Hall, 2001; Henschen, Statler, & Lidor, 2007). Coaches and athletes have indicated that they use imagery technique more than any other psychological technique. In addition, successful elite athletes use imagery in a more systematic and extensive way than less successful athletes (Morris, 2010). The effectiveness of imagery training in sport has been shown in the positive results of enhancing the athlete's performance and sport skills (Morris, Spittle, & Watt, 2005); enhancing self-confidence and motivation (Evans, Jones, & Mullen, 2004; Martin & Hall, 1995); and improving attention control (Calmels, Berthoumieus, & d'Arripe-Longueville, 2004). Athletes use imagery for different reasons, including skill learning and practice, strategy development and rehearsal, competition preparation, and coping with various sport stressors or obstacles (Morris et al., 2005; Vealey & Forlenza, 2015). Imagery is a technique that can be incorporated into many different mental training strategies and models.

Concentration techniques usually are associated with the ability to focus mental effort on what is most important in any situation while ignoring stress distractions. It is widely regarded as a vital determinant of athletic success (Moran, 2003). The concentrate skill can be improved by applying an appropriate PST program, including special concentration exercises, goal setting and self-talk strategies, preperformance routines, arousal control, and biofeedback training. Researchers have alleged that these techniques improve attentional skills and performance enhancement in different sports, such as shooting, archery, combat sport, tennis, and team sports (Moran, 2003). Some of the most prevalent causes of distracted attention and concentration are anxiety, worry, and irrelevant thoughts (Essig, Janelle, Borgo, & Koester, 2014). Concentration skills/techniques are probably the most important among all psychological skills for athletic performance but can be mastered only if the athlete has first learned to control his/her anxiety and arousal level through skills of relaxation (Henschen et al., 2007).

Self-talk is a psychological technique in which athletes talk to themselves—either silently or aloud and inherently or strategically—in order to stimulate, direct, react to, and evaluate events and actions. The use of self-talk is primarily focused on a verbal dialogue providing self-instructions or reinforcement (Hardy & Zourbanos, 2016; Hatzigeorgiadis et al., 2014). Research indicates that different types of self-talk (i.e., instructional vs. motivational) may be effective in enhancing different types of sport performance (Vealey, 2007). Highly-skilled athletes use self-talk in a more planned and consistent manner than less-skilled athletes, who tend to think reactively (Hardy, Hall, & Hardy, 2004). Several mental training techniques that are associated with self-talk, such as thought stopping, thought replacement, countering, reframing, and cognitive restructuring, are used in multimodal mental training interventions such as P^3 Thinking (Vealey, 2005) and energy management (Hanton & Jones, 1999).

Goal setting is a popular technique used to increase athletic motivation and enhance confidence, and as an effective tool for behavioral change. In applied sport psychology, three types of goals are discussed: outcome goals, performance goals, and process goals (Burton, Naylor, & Holliday, 2001). Outcome goals usually focus on the end result, for example, winning a medal or achieving specific ranking. These goals are often used for enhancing the athlete's motivation over a long period of physical and mental preparation. Performance goals help athletes focus on improving their own performance; they are more flexible, clear, and controllable. Examples of an athlete's performance goals are improving his/her percentage shooting in basketball and improving his/her personal best time in swimming. Process goals focus on the process of performing rather than on the end result of performance, for example, the technique of how an athlete performs a specific skill. These goals are used mostly in training session so that the athletes can focus on specific task demands (Vealey, 2007; Weinberg & Butt, 2014). Research with athletes found that goals must be incorporated into systematic PST programs that allow the athletes to plan, set, focus on, and adjust their personal attitude in relation to their goals (Gould, 2015; Vealey, 2005, 2007).

Biofeedback training (BFBT) is a technique used for gaining control of self-regulation, based on information or feedback received from an athletes' body and mind (Blumenstein & Orbach, 2014). Following intensive BFBT, psychological skills become automatic reflexes. Initially, most research on BFBT in sport focused on evaluating the efficacy of BFBT in helping athletes regulate performance anxiety and arousal levels (Zaichkowsky & Fuchs, 1988). Later on, research focused on combining BFBT with other psychological techniques/strategies, such as relaxation, imagery, breathing, and self-talk. It was found that BFBT as part of an intervention package has a positive effect on the athlete's performance (Blumenstein, Bar-Eli, & Tenenbaum, 2002). In addition, research has shown an augmenting effect of BFB intervention on training transfer from practice to competition, especially in motor learning and in perfecting movement techniques (Issurin, 2013). Today BFBT is used in research and practical programs as an integral component of PST in a variety of sport disciplines, such as archery, soccer, swimming, and wind surfing. One example is a program that was developed for the Canadian National Short Track Speed Skating team over a 3-year period leading up to the Vancouver 2010 Olympic Games (Beauchamp et al., 2012). In addition, the W5SA (Wingate 5-Step Approach; Blumenstein, Bar-Eli, & Tenenbaum, 1997) and the LMA three-dimensional approach (Learning-Modification-Application; Blumenstein & Orbach, 2012b) were developed and used as part of PST based on the periodization principle of sport training. For the first time, these approaches integrated systematic PST within an athlete's training process.

The LMA approach is an innovative psychological skills program that is composed of three dimensions (learning, modification, and application) and seven stress distractions. This approach includes BFBT together with other psychological techniques as one intervention package within PST. The LMA is integrated into the athlete's preparation and is based on the periodization principle of sport training. In the **learning** stage, athletes acquire fundamental psychological techniques in a laboratory setting, namely in controlled and sterile conditions, in order to enable the athletes to learn the basic foundations of each strategy. In the **modification** stage, the athletes continue to develop psychological skills and learn to perform short psychological strategies quickly and precisely, under a variety of stress distractions in laboratory and training settings. Finally, in the **application** stage, athletes use psychological strategies as part of preperformance routines, and they are applied in actual practice and competitions. The uniqueness of the LMA is the emphasis on the ability to transfer the psychological skills learned in the laboratory to the field. These psychological strategies within PST are trained under gradual stress distractions, which allow the athlete to prepare for real competition situations.

Preperformance routine is a pattern of physical and mental actions that athlete use before performance. More specifically, a preperformance routine is learned and practiced in order to direct the athlete's attention, help regulate psychological and physiological responses to stress, and allow motor processes to run with minimal conscious interferences (Jackson, 2014). During this period the athlete puts together relevant psychological skills and strategies (i.e., relaxation, imagery, concentration, positive self-talk) with physical activities, which allow the athlete to achieve an optimal level of arousal/activation and finally performance best. A preperformance routine lasts a brief period of time, is specific to each athlete, combines a number of psychological skills, and places the mind in a condition where it is ready to allow the body to perform (Henschen, 2005). Researches have suggested that preperformance routines can focus an athlete's attention, helping him/her to concentrate on the relevant aspects of the task and block out distractions (Boutcher, 1992). In addition, preperformance routines help athletes regulate their emotions, thoughts, and behaviors so that they can minimize distractions and focus on concentration (Jackson, 2014). Finally, this routine can help trigger automatic performance, which can facilitate a task's accuracy, especially self-paced tasks (Cotterill, 2010; Lidor, 2007).

INTEGRATION OF PST WITHIN THE ATHLETE'S TRAINING PREPARATION

A typical training program according to the theory and methodology of sport training is composed of three main phases: **preparation**, **competition**, and **transition**. The **preparation phase** is composed of two subphases: general preparation (GP) and specific preparation (SP). The main goal of GP is to improve general athletic and physical abilities, while the main goal for SP is to enhance the abilities and skills required for a specific sport type. The main objective of the **competition phase** is to improve the athletes' motor and psychological abilities in as many competitions as possible so that peak performance can be achieved. Finally, the objective of the transition phase is to enhance physical and psychological rest and to make sure that an acceptable level of the athletes' GP is maintained. In each phase, four specific types of preparations are included: physical, technical, tactical, and psychological (see Bompa, 1999; Bompa & Haff, 2009; Carrera & Bompa, 2007; Holliday et al., 2008; Zatsiorsky, 1995). In this section the integration of psychological preparation within the athlete's training program will be discussed. For more details and additional information regarding these phases and the other three preparations, see Chapters 5, 6, and 9 of this book.

A planning tool for developing the athlete's training program is the periodization principle. When applying a psychological program, the main focus should be on the incorporation of the mental session within the athlete's training program, while linking it with the other preparations: the physical, technical, and tactical. Moreover, it is important to modify the mental sessions based on the volume and intensity of the athlete's physical training. The periodization principle adds a planning timeline tool to the athlete's overall preparations, and therefore is used as a guideline for PST training—especially in the LMA approach (Balague, 2000; Blumenstein et al., 2007; Blumenstein & Orbach, 2012b; Blumenstein & Orbach, 2018; Holliday et al., 2008). This approach is used as a framework that illustrates a system for the integration of PST within an athlete's training process.

In the **GP subphase** the athlete improves his/her general physical condition, including power, speed, flexibility, and endurance, and trains using a large number of low-intensity exercises and repetitions, monotonous loads, a difficult daily regime, a high volume of training with firm coach demands, and so forth. Accordingly, the main objective of PST in this period is to help the athletes cope effectively with the physical load they are exposed to during practice. Therefore, the sport psychology consultant has to focus during this period on two main goals. The first goal is to assist the athlete to recover after hard training, and the second one is to strengthen the athlete's sport motivation, confidence, and goals for the upcoming season. To achieve these, the psychological techniques applied in this period consist of relaxation (20–25 min), special music sessions (15–20 min or shorter duration), imagery featuring nature pictures, video clips (10–15 min), and biofeedback training as well as combinations of these techniques, such as relaxation and imagery, relaxation and music, and relaxation and biofeedback (see Blumenstein et al., 2002; Blumenstein et al., 2007; Blumenstein & Orbach, 2012a; Dosil, 2006; Statler & Henschen, 2009; Vealey, 2007).

From the perspective of the LMA approach, in the **learning stage** athletes acquire fundamental psychological techniques in a laboratory setting, namely in controlled and sterile conditions, so that they may learn the basic foundations of each strategy. Moreover, toward the end of this stage, the athletes train under light stresses distractions (i.e., verbal positive and negative motivation) (see Table 3.1).

Table 3.1 Recommended psychological techniques for general preparation phase according to the LMA approach, including examples and their objectives

Phase	LMA stage and stressors	Psychological technique	Objectives	Application, duration, place
General preparation	Learning stage with light stress distraction	Relaxation	Mental recovery	20–25 min–end of the week 10–15 min, 1–2 times/week 5–10 min, 3–-4 times/week
		Imagery (nature pictures)	Mental recovery	10–15 min, 1–2 times/week
		Music (relaxation oriented)	Mental recovery	15–20 & 5–10 min, 1–2 times/week
		Biofeedback (training & games)	Mental recovery	10–15 min, 1–2 times/week
		Goal Setting	Self-confidence, *↑motivation	10–15 min, 1 time/week 5 min, 1–2 times/week
		Combinations: Relaxation & Imagery Relaxation & Music Relaxation & Biofeedback	Mental recovery	20–25 min at laboratory settings, 1–2 times/week

*↑ improved/increased.

In the **SP subphase** the goal is to further develop the athlete's physical abilities according to the unique physical and physiological characteristics of a specific sport (Blumenstein et al., 2007; Blumenstein & Orbach, 2012a,b; Bompa & Haff, 2009). Accordingly, the intensity of practice increases and the number of repeated exercises substantially decreases. The main objective of PST within SP is to practice psychological skills and strategies in order to enhance the physical, technical, and tactical skills required for a specific sport type. The athlete is exposed to a variety of actual environmental factors related to a competition situation. Therefore, the psychological techniques are performed under special stress distractions and are related to sport-specific demands. In addition, during this period the athlete participates in some competitions, and therefore the PST in the individual sport should focus more on parameters such as self-confidence, self-regulation, and concentration. In team sports, the PST centers on developing team cohesion, communication, leadership, group dynamics, and relationships between players. The sport psychology consultant takes part in athlete/team practice, introduces psychological techniques in training, identifies weaknesses in the athlete/team psychological preparation, and, finally, optimizes the precompetitive routine. In this period it is important to pay attention to the interaction between the psychological, tactical, and technical preparations. It is especially important to consider the *technical preparation*, in which mental and motor control play a vital role.

From the perspective of the LMA approach, in the **modification stage** the athlete/team is being prepared for real-life situations. Therefore, the athletes continue to develop their psychological skills, performing psychological strategies shortly, quickly, and precisely, under moderate stress distractions (i.e., specific demands, reward/punishment). The skills are adapted in line with technical-tactical purposes and the sport's requirements. For example, the athletes will focus on concentration and imagery techniques, in which they visualize technical elements of themselves and their opponents in competitive situations. The mental practice is provided initially in the lab and later in the training setting (Blumenstein et al., 2005; Blumenstein & Orbach, 2012b, 2014) (see Table 3.2).

Table 3.2 Recommended psychological techniques for specific preparation phase according to the LMA approach, including examples and their objectives

Phase	LMA stage and stressors	Psychological technique	Objectives	Application, duration, place
Specific preparation	Modification stage with moderate stress distraction	Relaxation	Focusing attention & arousal regulation	1–5 min/training settings 5–10 sec/practice between exercises, fights, games
		Imagery	Mental readiness, focusing on aspects of technical-tactical performance	1–3 min/practice; warm-up, between exercises, fights, games
		Self-talking	Mental readiness, self-confidence	10–15 sec/before exercises, fights, games/warm-ups
		Concentration exercises	Preparation for performance; mental readiness	5–10 min/lab settings
		Breathing exercises	Mental readiness, mental recovery	1–2 min/training; before and after performance
		Biofeedback training	Mental readiness, self-confidence	10–30 sec/laboratory settings. 1–2 times/week

In the **competitive phase**, the intensity of the performed technical elements increases, and the repetitions decrease, while the total time of training decreases. In addition, the athlete intensively participates in a variety of competitions. In individual sport the athlete takes part in several competitions that prepare him or her for the main target competition, such as National, European, and World Championships, or the Olympic Games. In team sports the athletes usually compete in national leagues, in which games are on a weekly basis for a period of several months.

Therefore, the main training goals in this period are the following:

(1) To maintain general and specific physical conditions

(2) To improve the athlete's techniques and tactics

(3) To cope with competition stress

(4) To gain competitive experience

(5) To improve team communication, interaction, and team tactics

The ultimate goal of PST in this period is to transfer most of the mental training from the laboratory to the field and to apply the psychological strategies in different training and competition situations. Moreover, the psychological strategies are modified to the training and competitions demands, such as the duration and specificity of the sport. Therefore, in this period psychological skills are more related to the kind of sport, the actual environment factor of competition, and being an integral part of the precompetitive routine (Blumenstein et al., 2007; Blumenstein & Orbach, 2012a,b).

From the perspective of the LMA approach, in the **applied stage** the mental strategies are part of the precompetitive and preperformance routines and are applied in actual practice sessions and competitions where the athletes are exposed to more authentic situations and real-life distractions. The athlete is able to apply the psychological strategies quickly and effectively in training and competition. Moreover, the strategies are integrated with the technical, tactical, and physical demands of competition and sport. Taking the above into account, during mental training sessions in this stage, the athlete practices psychological techniques such as concentration and imagery, accompanied by highly stressful competitive distractions (e.g., the use of recorded competition films and noises) (Blumenstein & Orbach, 2012b, 2014) (see Table 3.3).

Table 3.3 Recommended psychological techniques for competitive phase according to the LMA approach, including examples and their

objectives

Phase	LMA stage and stressors	Psychological technique	Objectives	Application, duration, place
Competition	Application stage with high stress distraction	Relaxation & music; relaxation & imagery	Mental readiness, mental recovery	5–10 min; weekend or during competition settings, between exercises, fights, practices
		Imagery	Mental readiness, focusing on aspects of technical-tactical performance	1.30 min—rhythmic gymnastics 4 min—judo Swim race—swimming distance w/time Basketball—combinations Rowing, kayak—race distance
		Self-talking	Mental readiness, focusing on aspects of technical-tactical performance	10–15 sec/competition and training settings, before performance or analysis of earlier competitions
		Concentration exercise	Preparation for performance; mental readiness	5–10 sec/competition and training settings, before performance
		Breathing exercise	Arousal regulation	10–15 sec/competition and training settings, before performance
		Precompetition mental readiness	Mental readiness	competition settings; before performance
		Biofeedback training	*↑Concentration and self-confidence	5–10 min/lab settings: 1–2 times/week

*↑ improved/increased.

In the **transition phase** the goal is to enable the athlete/team to take part in physical and psychological rest and recovery. Athletes are advised to remain active during this phase so that they will be better prepared for the next preparation phase. The main objective of PST within the transition phase is mental recovery, such as relaxation, listening to special individualized music, and breathing exercises while incorporating BFB games (Blumenstein & Orbach, 2012a, 2012b, 2014) (see Table 3.4).

Table 3.4 Recommended psychological techniques for transition phase, including examples and their objectives

Phase	Psychological technique	Objectives	Application, duration, place
Transition	Relaxation	Mental recovery	10–15 min; 1 time/week/lab settings or home
	Music	Mental recovery	10–15 min; 1 time/week/lab settings
	Biofeedback games	Mental recovery; self-confidence	10–15 min; 1 time/week/lab settings

SUMMARY

Psychological preparation is an integral and significant part of athlete/team preparation in modern sport. This chapter focuses on the fundamentals of psychological preparation, psychological skills, psychological strategies, and the integration of PST within training process. The uniqueness of the integration of PST with an athlete's training process is demonstrated by the framework of the LMA approach. Based on the theory and methodology of sport training, the periodization principle is a basic fundamental method for the annual training process. Periodization of physical training is both a well-established method of long-term athlete development and a competitive preparation tool within the physical training domain. In this chapter, the focus is on using the periodization principle as a tool for the development of PST and the outline of LMA, as a way of integrating PST within the training process.

Moreover, psychological preparation is an integral component of all three training preparation phases (i.e., preparatory, competitive, transition), and its elements strengthen other preparation components (i.e., the physical, technical, and tactical). Each training session and practice should include a psychological ingredient or component. This holds true for all active coaches and athletes on every training day. This chapter focuses on the integration of psychological preparation within the athlete's training process. The psychological factors appropriate for physical, technical, and tactical preparation are discussed in other chapters of this book.

Chapter 4

APPLICATION OF INTEGRATED PERIODIZATION

Tudor Bompa, PhD / Boris Blumenstein, PhD / Iris Orbach, PhD / James Hoffmann, PhD

While many sports professionals and athletes are exposed to many aspects of modern training, periodization and most training programs are still not integrated into a complete whole. At the same time, the present level of training and research findings in related sciences of sports evolve so rapidly that a team approach becomes a necessity. Specifically, this means to offer to athletes and teams the necessary services to reach maximum potential.

Equally important for the sports professionals, and especially for coaches and athletes, is to continually test and monitor athletes rate of development and to constantly analyze and offer services in the area of the methodology of developing specific motor abilities, nutrition, and sports psychology.

Access to specialized physiological monitoring/evaluation is also a requirement for athletes whose training objectives require a high level of performance. As such, the proposed concept of Integrated Periodization means to combine all aspects of training and to match them during training, prior to and during competitions.

Of great importance is maintaining the same professional services during the *transition* phase, where psychological relaxation, and nutrition plans are necessary to facilitate recovery and regeneration after a long year of training and competitions. This phase of recovery and relaxation is a necessity prior to a new beginning. Before a new preparatory and competitive phase starts, where pain and strain are often the norm of an athlete's life. Neglecting to use such services could compromise recovery and regeneration by starting a new annual plan with residual fatigue, not exactly in the best physiological and psychological conditions.

For maximum benefit, integrated periodization is suggested for these:

1. The annual plan
2. All phases of training inside the annual plan
3. As is relates to the methodology of developing the dominant motor abilities in the selected sport (main abilities necessary to maximize performance in a given sport)

DEVELOPING AN ANNUAL PLAN

The annual plan is the physical manifestation of our integrated periodization concept. It is the culminating tool of principled training, nutrition, and psychological practices, which moves from theory into real-world application. The annual plan is a detailed map describing all the events occurring throughout the year, relating to the athlete's training and competition schedule. It is an essential component of a periodized plan because it begins to divide the training year into smaller distinct phases each with specific objectives. This ensures that athletes are peaked at appropriate times for competition, while still retaining the capacity to train hard and recover throughout the year.

The annual plan is driven by the needs analysis of the athlete and is specifically tailored to short- and long-term goals. An annual plan can be as simple as one competitive season worth of training or multiple annual plans linked together, providing a road map for years of training. This degree of planning may seem daunting or tedious to coaches but may be essential for long-term sport success. Using an annual plan can help deal with obstacles to continued progress, both planned and unplanned, to ensure the athlete is on track with preplanned benchmarks.

The annual plan consists of anything relating to an athlete's competition, training, or recovery schedule. This can include (but is not limited to) the following:

- Sport practice
- Physical training
- Psychological training
- Body composition alteration and bodyweight management
- Fatigue management and recovery strategies
- Travel, altitude adjustment, and jet lag
- Holidays (personal, national, religious, etc.)
- Competitions
- Testing and monitoring
- Nutrition
- Promotional events

Though this amount of information may initially seem overwhelming, it is generally easier to plan around these issues rather than dealing with them unexpectedly. Not every plan is perfect, and inevitably complications will arise. Those with a well-developed annual plan will be better suited to deal with adversity and stay on target.

PRELIMINARY PLANNING

To begin developing an annual plan, there is a hierarchy of steps that should be followed to minimize back tracking and overcomplication. The first step is to simply find a medium in which you can work on the plan and the athlete can follow. This can be as simple as a notebook, spreadsheet, or calendar application, or upwards of expensive specific coaching or team monitoring software. All have their merits in terms of utility and practicality, but ultimately you must decide which is the best for you. A good place to start is using a simple spreadsheet.

The second step involves establishing a timeline. This can be done by mapping out days, weeks, or months depending on the users' preference. Generally, it is recommended to start on a large scale and progressively work down to smaller units of time. For example, instead of starting by planning out individual sessions, one should start mapping out the annual plan's entire year, then mapping out each month, then each week, then each day, and lastly each session within each day. A very simple spreadsheet approach would be mapping out all 52 weeks throughout the year and then use linked tabs to break down the specific dates of each macrocycle. Once the macrocycles are labeled, the microcycles can be labeled as well. The chosen application and process should suit your own personal preferences; something that you and your colleagues can open every day and feel comfortable viewing and editing.

Once the timeline has been labeled, the third step in developing the annual plan is detailing all of the competition dates. This can include all preseason, early season, late season, championship, scrimmage, play-off, tournaments, and friendly matches currently known. This will also include any pertinent information regarding travel, time of day, equipment, and location. Although all of the competitions may not require specific tapering or fatigue management protocols, the competitions serve as critical time points when strategizing the phasic approach throughout the year.

Next will be to enter in all holidays and mandatory down times for the athlete. This should include compliance with NCAA or any governing body for sport, religious holidays, national holidays, personal holidays, sponsorship appearances, public service, or any day the athlete would be unavailable for training, practice, or dedicated recovery time. This may seem a bit odd, but there can be a surprising number of external restrictions where coach-athlete interactions are very limited. Therefore, mapping out periods where the athlete is essentially unavailable for specific training or recovery practices can be extremely helpful.

The next portion of planning involves mapping out traveling periods. This includes the more apparent time spent actually in transit but can also include time needed to adjust for things like jet lag and altitude. For local competitions this can often be a nonissue; however, for teams that travel substantially, be it domestic or international, this can become problematic. Several collective days may be required to travel to individual competitions throughout the year, and that is generally time that cannot be spent training or doing specific recovery activities. Additionally, coaches may need to seek out facilities for physical training or sport practice during this transit time and plan accordingly. International travel may require arriving to the destination weeks in advance to acclimatize the athlete to the new time zone and potential altitude adjustment. Since training and recovery will be constricted during these periods, they should be preemptively factored into the plan.

The last portion of planning before training is outlined involves mandatory testing sessions. This can include NCAA testing, USADA and drug testing, laboratory testing and monitoring, and regular checkups with their physicians or sports medicine team. This is of particular importance for athletes who require regular monitoring of blood work or existing injuries, as this time came significantly accumulate throughout the year (rightfully so). Each governing body for sport tends to have slightly differing regulations on medical and drug testing, in addition to the need for internal performance testing and monitoring by the coaching staff. Some medical and drug testing can take considerable time to perform, and in some cases to recover from, so they should be accounted for within the annual plan.

PROGRAMMING

Once the preliminary factors and obstacles have been outlined, coaches and athletes can begin the programming portion of the annual plan. Generally, this is the most enjoyable and rewarding portion of this process; however, there is a proper order of operations that coaches and athletes should be aware of before beginning. A common problem many experience at this stage is anxiousness to begin putting ideas down into writing. Although the enthusiasm is well-placed, several critical and common mistakes often result:

1. Failing to reference or generate the needs analysis of the athlete
2. Starting with preparatory periods rather than competitive or transition periods
3. Starting on the session to session, or microcycle strategies, and working to larger timescale strategies

Luckily all these problems can be avoided or remedied simply by following a structured order of operations. Although you may tailor these steps to your own individual needs, the authors recommend following these steps until you develop the system that works best for you.

REFERENCE THE NEEDS ANALYSIS

The needs analysis of the athlete will play a huge role in determining not only the physical characteristics of training, but also all the integrated approaches previously mentioned. Before the programming of any exercises, skills and techniques, tactics, body composition alteration, or psychological skill enhancement we first must understand the strengths and weaknesses of the athlete in their current state and how they compare to their peers. Here, the use of normative standards, game statistics, testing and monitoring, and the coach and athlete feedback are critical to developing short- and long-term training strategies. These strategies will very likely differ from person to person, even within the same sport or team, as well as within the individual across different time points throughout their career and competitive season. Developing an annual plan without a needs analysis is akin to planning a trip without a destination.

BEGIN PROGRAMMING THE COMPETITIVE PHASES, AND THEN WORK AWAY OR "BACKWARDS" TOWARD PREPARATORY PHASES

Competitions serve as critical time points in the generation of an annual plan. If developing an annual plan could be compared to building a house, the needs analysis would serve as the blueprints, and the competitive periods would serve as the foundation. It's nearly impossible to effectively install plumbing, electrical, appliances, roofing, and other aspects of a home to a structure without a foundation. Likewise, when developing an annual plan, it becomes incredibly tedious and frustrating to develop phase potentiated strategies when starting from random time points. When starting from the competitive phases, it becomes immediately apparent where the additional phases and subphases must be placed.

One of the first things that can be addressed starting from competition are fatigue management strategies such as peaking and tapering, deloading, or active rest, which lead into or follow the competition itself. Once tapering periods are outlined, then specific preparatory leading into the taper can be developed, with goals and durations dictated by the needs analysis. Then finally the general preparatory or "off-season" phases can fill the remaining time between other competitive phases or transition periods. Similarly, if we looked at that same competition and worked our way in the other direction (forward in time), we can immediately begin mapping transitional or active rest phases following that competition. We can also program what is being done between competitions for sports with a season or what is done leading up into postseasons, championships, or any subsequent competitions.

The order from competitive phases to preparatory phases allows coaches and athletes to use time as more of a finite resource, rather than a guessing game or puzzle. Instead of having a blank slate of 15 weeks to prepare for competition, coaches can begin allocating that time into smaller and more manageable doses to phase potentiate training and recovery strategies. Attempting to work forward in time to competitions often becomes a tedious "guess and check" approach, which often requires numerous revisions and overhauls. Starting at competitions and working away or "backwards" in time can help prevent this frustrating process.

START WITH LARGE STRATEGIES AND WORK DOWN TO SMALLER STRATEGIES

From our competitions, the first strategy should be to begin marking all the large-scale phases of the annual plan. This includes the competitive phases, preparatory phases, and transitional phases. The next step will be to divide each major phase into its component subphases when applicable. This can include the general and specific preparatory periods, preseason phases, active phases, and others dictated by the needs analysis. Once each subphase has been outlined, it becomes increasingly easy to develop month to month strategies, week to week strategies, day to day strategies, and lastly session to session strategies.

If we go back to our house building analogy, trying to develop session to session strategies without knowing where they fit in the competition timeline is akin to choosing a paint color for a wall that has not been built yet. Although it's possible, the decision is mostly arbitrary, and you might find it doesn't make much sense after seeing the completed structure. Changing a paint color is also relatively easy, whereas trying to change a wall is more challenging. Once the overarching structure of training has been developed, session to session strategies become more "fine-tuning." They are more malleable and flexible based on the individual needs and concurrent state of the athlete. In general, a systematic approach would involve mapping out the competitive periods first, followed by preparatory and transitional phases, any subphases within each phase, and then working from months to weeks to days to sessions within each subphase.

INTEGRATED PERIODIZATION PLAN

A basic model for an annual plan is suggested by figure 4.1 for a speed-power dominant sport (Bompa, 1999). Specific training phases are defined, as it relates to the competitive calendar. Since this annual plan is made for a sprinter in track and field, periodization of motor abilities is suggested only for speed and strength. Finally, since our scope is to exemplify an integrated periodization, our example also refers to specific psychological and nutrition plans that were used in those years. Since the competition calendar and the dominant abilities differ from sport to sport, several examples of annual plans are exemplified in chapter 5 (figures 5.1–5.11).

A typical training program is composed of three training phases: preparation, competition, and transition (Bompa, 1999). Throughout the training cycle, the objective of physical preparation is to develop the relevant motor abilities of strength, speed, and endurance. Nutritional and psychological interventions may be integrated into the regular training schedule from the first session of the preparatory period throughout the competition and transition phases.

Months	1	2	3	4	5	6	7	8	9	10	11	12
Training Phases	Preparatory					Competitive						Transition
Subphases	Preparatory		Specific		Precompetitive	Official competitions					Unloading	Transition
Periodization: Speed	Anaerobic and aerobic endurance	- Maximum speed - Anaerobic endurance		- Maximum speed - Specific speed - Agility - Reactive agility	- Sport-specific preparations - Specific speed - Agility - Reactive agility						Unloading	Play, fun, rest
Periodization: Strength	Anatomical adaptation	Maximal strength		Power	Maximal strength	Conversion of power		Maintenance of power or maximal strength				Compensation
Periodization: Psychological Preparation	- Evaluate mental skills - Strength sport motivation and goals - Learn and practice basic psychological strategies/ techniques		- Modify psychological strategies/ techniques according to sport demands and the athlete's qualities		- Transfer psychological skills from laboratory to field - Develop and apply psychological skills/strategies within preperformance routine and the athlete's performance							- Recovery techniques - Active rest
Periodization: Nutrition	- High carbohydrate - Moderate protein	- High protein - Moderate carbohydrate	- High carbohydrate	- High carbohydrate - Moderate protein	- High carbohydrate - Moderate protein		Fluctuates according to the competitive schedule				High carbo-hydrate	Balance diet

Figure 4.1 A basic model for an integrated periodization annual plan

NUTRITION INTEGRATION ACCORDING TO TRAINING PHASES

The role of nutrition and training is inherently linked to performance. Nutritional interventions to elicit training adaptations should follow a phase-based approach. Since each phase of training has different goals and varied fatigue accumulation, nutrition recommendations change accordingly. Nutrient timing, type, and quantity take a major role in physical training and psychological success. Although physiological adaptations and nutritional needs for various sports are different, some commonalities are found across sporting disciplines: (1) Energy, fluid, vitamin, mineral, and macronutrient requirements must be met (Sahlin, 2014). (2) Repair and maintenance of tissues is dependent on tissue type, remodeling time, hormones, and nutrient availability and timing (Houtkooper, Abbot, & Nimmo, 2007; Tortora & Derrickson, 2012). (3) Buffering capacity of the muscles can be improved through supplementation and training (Sahlin, 2014). (4) Different forms of training can turn on cellular signaling pathways for protein-specific adaptations, which ultimately rely on nutrition or expense of tissue (Spriet, 2014). (5) Acute and chronic adaptations to training are facilitated by proper and consistent nutrition strategies (Spriet, 2014). (6) Studies have indicated DNA oxidation induced by training may be attenuated by supplementation with antioxidants (Ziv & Lidor, 2009).

The needs to integrate nutrition and training are not entirely bound to physiological adaptation. Some practical needs include proper body composition for a specific sport or position, performance improvement, increased knowledge base of the athlete, and individualization. Extreme physical demands and weight classed competition requires lean body mass with minimal body fat. For example, a study of 45 professional Rugby players following a relatively low carbohydrate (CHO) and high protein diet showed considerable and significant decrease in the sum of seven-skinfold-site analysis from 93.2 (25.9) (preseason) to 87.6 (25.8) (postseason) for forwards over a 10-week period (Bradley, et al., 2015). In addition, Bradley and colleagues (2015) note significant improvement in all measures of strength (squat, RDL, bench press, weighted chins) over the entire preseason.

The nutritional knowledge base of the athlete has an immediate effect on consistency and adherence to nutritional principles. Jacobsen and colleagues (2001) found that knowledge base of nutrition and incorporation of principles into the training regimens of college varsity athletes has increased modestly. This was attributed to increased delivery of nutritional knowledge from strength and conditioning coaches (Jacobsen, Sobonya, & Ransone, 2001). Each athlete carries different genetics, somatotypes, and tastes, which require a certain degree of individualization. Integration of nutrition and sports training will ultimately reveal a situation that allows for situational modification and individualization.

Integrating nutritional interventions into the annual plan becomes streamlined once all the preparatory, competitive, and transition phases have been established throughout the training cycle. Each training phase pairs with a synergistic nutritional phase based on the needs analysis of the athlete, their competitive schedule, and their current physical and psychological state. These interrelations are summarized in figure 4.2.

Training Phase	Preparatory		Competitive	Transition
	General	Specific		
Training Goals	Developing sport and training skills	Mastery of training and sport skills	Alleviating fatigue and elevating preparedness	Link multiple competitive cycles together
	Elevate work capacity and general fitness	Elevate sport-specific strength, explosiveness, and conditioning	Continued improvement or maintenance of fitness	Alleviate physical and psychological fatigue
	Basic sport tactics	Team and individual tactics	Perfecting skills and tactics	Maintain basic fitness
Training Phase	Body Composition Alteration	Stabilization	Recovery and Performance	Active Rest
Nutritional Goals	Gain muscle or lose fat	Achieve and maintain within 2% of competition bodyweight	Shift macronutrient priorities to carbohydrate	Reduce stress of dieting
	Develop basic diet and hydration routines	Refinement of eating and hydration patterns	Maintain euhydration	Maintain a reasonable bodyweight
	Consistent monitoring of bodyweight	Fatigue alleviation	Maintain lean body mass	Maintain lean body mass

Figure 4.2 Nutritional integration within training phases

Major body compositional change should occur primarily during general preparatory periods. The needs analysis should outline the body compositional standards the athlete needs to achieve in order to potentiate both short- and long-term goals. This could mean gaining muscle, losing fat, or simply developing fundamental feeding and hydration routines. If body compositional change is not a priority during this time, emphasis should be placed on developing successful daily nutrition routines, consistent monitoring of bodyweight, and developing a large work capacity through overload training.

As the athlete transitions into specific preparatory periods, major body compositional changes should cease, and the athlete's bodyweight should be stabilized within about 2% of ideal competition weight. This stability is achieved by adjusting calories to isocaloric levels. Fundamental feeding and hydration routines developed in the previous phase should continue to be refined while maintaining stable weight. Attempts to continue body compositional change at this stage is counterproductive to the athlete due to these:

1. The accumulated fatigue generated by the high volumes of training and dietary changes needed to drive body compositional alteration can mask and prevent the expression of more sensitive fitness characteristics such as maximal strength, power, and speed.

2. The athlete may fail to develop the "feel" or mind-body connection needed to express high levels of per-

formance at their competition bodyweight if weight is still changing during this stage.

3. The training volumes during specific preparatory periods are sufficient for muscle retention under isocaloric conditions; however, they are likely too low to stimulate growth and too low to retain muscle under hypocaloric conditions.

4. Hypo- or hypercaloric conditions and/or high-volume training during this stage disrupt phase-potentiated progression of training and impede appropriately timed peaking for competition.

Once the competitive season or period begins, the athlete should make a subtle shift in nutrition to favor recovery and performance. This is done by placing a high macronutrient priority on carbohydrate consumption to ensure substrate-level recovery from intense training and competitions. The feeding and hydration routines should largely be set at this point, but athletes may individualize portions and food types to meet their competition and training specific preferences. Specialized routines such as weight cutting, meals between multiple competitions within the same day, and within competition nutrition, should be rehearsed during lesser competitions before moving onto the priority competitions.

Once the major competitions are completed, the athlete should begin a temporary active rest phase. The guidelines for active rest phases apply to nutrition as well as training. Food intake should be less measured and less structured as the athletes recover and transition from one annual plan or competition schedule to the next. This is meant to be complementary to training active rest phases in reducing the mental and physical stressors of vigorous and diligent training and nutrition. The athlete is given a break from their normal routine and allowed to largely eat what they would like so long as they maintain a few simple conditions:

- Maintain a reasonable bodyweight relative to their ideal competition weight.
- Maintain lean body mass.
- Maintain euhydration.

So long as these simple conditions are met, the athlete can take a well-deserved break from the rigors of dieting to cultivate desire for future training cycles.

PSYCHOLOGICAL INTEGRATION ACCORDING TO MOTOR ABILITIES

The psychological facet of sport performance has been recognized for decades in theory and practice. However, the implementation across training phases is often relegated in support of other scientific disciplines such as biomechanics and kinesiology. The reluctance to integrate is rooted in the notion that psychological strategies contain a certain subjectivity, whereas natural sciences have objective measurement approaches (Smith, 1989). To the contrary a multitude of studies have established objective measures that solidify the connection between psychological techniques and altered physiological outcomes. Tod and colleagues (2005) established a significant increase in maximal force production when psychological preparation was applied before a lift. In another study, muscular endurance in the bench press and squat was found to increase when participants were instructed to mentally change to an external focus (Marchant, Grieg, Bullough, & Hitchen, 2011).

The needs of integrating psychology with training start with complete preparation of the athlete (Balague, 2000; Blumenstein, Lidor, & Tenenbaum, 2007; Blumenstein & Orbach, 2018; Holliday et al., 2008).

Psychological interventions should be integrated into the regular training schedule from the first session of the preparatory period. The comeback to regular training after the transition phase is a challenging task for the athlete. In order to achieve a positive start of the upcoming season, the athlete is required for high motivation, determination, ability to relax and concentrate, clear goals, and self-confidence.

Strength is a fundamental motor ability for all sports. There are some concerns regarding **strength** development that should be considered. A major goal of strength training is to maintain performance quality and optimize training return. To achieve that, the athlete should develop awareness to training method and exercise quality, which is a key psychological concern for effective training (Collins & MacPherson, 2007). Another concern is the difficulty of showing the direct relationship between specific strength training exercise and the athlete's sport-specific movements, meaning the transfer of the strength exercise to the actual athlete's performance. Therefore, the athlete should be educated regarding the effect of strength training on the athlete's performance, especially during the preparatory phase. To achieve that, the athlete should mentally go through the sport-specific action ("mental simulation of movement"—MSM), followed by the appropriate strength exercise, in order for positive transfer (Smith, Collins, & Holmes, 2003). In addition, the athlete should acquire psychological techniques such as motivation, concentration, muscle relaxation, and self-talk for effective transfer from mind to movement, especially during the preparatory and competitive phases.

A special attention was given to a few psychological techniques in relation to the motor ability of muscular strength. One of the psychological techniques is an appropriate and effective self-regulation of emotion (Collins, 1999). This means that the athlete should be able to identify the most appropriate emotional state for optimal training sessions and injury prevention (Collins & MacPherson, 2007). In addition, imagery has produced positive effects on individual parameters of performance such as muscular growth (Lebon, Collet, & Guillot, 2010; Reiser, Busch, & Munzert, 2011). The more vividly subjects imagined contracting their muscle, the more strength they gained. Moreover, kinesthetic imagery produced significant effects on muscular strength but only when subjects simulated the feeling of performing the movement from inside the body (Yao, Ranganathan, Allexandre, Siemionow, & Yue, 2013). Some other studies have indicated the positive effect of mental imagery on the development of strength specific to sport discipline such as combat sport (karate; Fontani et al., 2007). The motor imagery training strengthened the brain-to-muscle command (mind-muscle connection), which improves the recruitment of motor units. This led to higher muscle output, which resulted in greater measured strength. Important to note the relevance of the biofeedback training as a tool for effective muscle strength. For example, Croce (1986) found that a combined training program, which consisted of an isokinetic exercise and EMG BFB (electromyography and biofeedback) training, was effective for increasing muscle strength. The above psychological techniques have been used during the beginning of the preparatory phase mainly as independent techniques. However, toward the end of the preparatory phase and during the competitive phase, the psychological techniques are being used as an intervention package accompanied with biofeedback training (Blumenstein & Orbach, 2014).

It is a well-known fact that relaxation is essential for **speed** performance (Sheard & Golby, 2006). Athletes can move fast and change directions best when he or she is relatively relaxed. From a practical perspective, trying too hard, meaning becoming stiff and forcing one's muscles, is invariably counterproductive. Research has produced positive effects of psychological strategies on athlete's speed (Sheard & Golby, 2006). Major strategies/techniques that have been used in research are muscle relaxation, imagery, breathing, and biofeedback training together with relaxation and imagery. Similarly to research findings, sport psychology practitioners apply in their work relaxation and imagery of speed performance (e.g., 100 m run or swim) with an objective measure for evaluating time during imagery (e.g., run/swim). During the preparatory phase, imagery training is considered an important strategy, as well as during the competitive phase, in which performance time during imagery is being used for improving speed (Blumenstein & Orbach, 2012a,b; see details in chapters 3 and 5). In addition, biofeedback training has positive effect on speed performance. This effect was documented in research as well as in practical work (Bar-Eli & Blumenstein, 2004; Blumenstein, Bar-Eli, & Tenenbaum, 1995).

There appears to be a lack of studies examining the efficacy of psychological interventions on competitive **endurance**-based athletic performance. Research has indicated that in order to achieve efficient *endurance*, training an athlete should use specific psychological strategies/techniques (Thelwell & Greenlees, 2003). Depending on the goal of the training, athletes use either association or disassociation styles. In association style the athlete may focus on task-relevant components such as breathing, and in disassociation style the athlete may focus on music and surrounding scenery (Antonini-Philippe, Reynes, & Bruant, 2003; Wrisberg & Pein, 1990). For example, for duration and distance goals, disassociation style is beneficial with the cost of slower progress and lower level of performance. On the other hand, when the goal is to emphasize performance in order to optimize achievement, association style is the recommended one (Collins & MacPherson, 2007). In addition, psychological packages that include goal setting, imagery, self-talk, and relaxation can enhance endurance performance on high-level endurance athletes (e.g., triathlon and running; Patrick & Hrycaiko, 1998) and on lower-level performers (Thelwell & Greenlees, 2001).

From periodization perspective, we recommend emphasizing disassociation style in preparatory phase, while focusing more on association style in competitive phase. In regard to psychological techniques, the trend will be using a variety of techniques such as relaxation, self-talk, imagery, and goal setting by itself during the preparatory phase, while combining the techniques within one intervention package during competitive phase according to sport discipline, athlete's current state and characteristics.

To summarize the above, psychological intervention is an important factor during the athlete's training program. In *general preparation* psychological skill training (PST) is oriented on strengthening the athlete's sport motivation and goals for the upcoming season. In laboratory setting, athlete learns and practices basic psychological techniques/strategies such as relaxation, imagery, self-talk, biofeedback training, and self-confidence. In *specific preparation*, most of psychological strategies/skills are modified accordingly to sport demands and the athlete's psychological qualities. In this phase, the focus of PST is on developing and practicing mainly relaxation, concentration, self-confidence, self-regulation, and imagery. In team sports, PST includes also team cohesion, communication, and leadership. In *competition phase*, the focus of PST is on transferring learning skills from the laboratory to field and applying the psychological strategies/skills within preperformance and pregame routines to attain optimal state. In addition, the application of the preperformance routine is being done gradually from minor competitions to major ones. In the *transition phase*, the focus of PST is on recovery techniques such as muscle relaxation, autogenic training, relaxing breathing, and emotional management. This phase of recovery and relaxation is a necessity prior to a new beginning. Before a new preparatory and competitive phase starts, where pain and strain is often the norm of an athlete's life. Neglecting to use such services could compromise recovery and regeneration and start a new annual plan with some residual fatigue, not exactly in the best physiological and psychological conditions. In the following chapter in Figures 5.1–5.11, the general concept of the psychological strategies according to the periodization principle in different sport disciplines is presented. However, the PST should be individualized based on the unique objective (e.g., age, gender) and subjective (e.g., skill level, emotional state) characteristics of the athlete's current state and its sport discipline.

SUMMARY

Although training, nutrition, and psychology are all incredibly valued in their own respective ways, the training portion is generally the most straightforward to outline first. The systematic nature of periodized training makes it the driving force behind an integrated annual plan. Nutritional and psychological strategies are generally supplemental to the training strategies, meaning they support the process rather than guide it. Thus, nutritional and psychological phases will generally be categorized and paired with complementary training phases. Trying to adapt training to nutrition or psychology phases tends to be more problematic. For example, planning hypercaloric phases to gain lean body mass without corresponding hypertrophy training generally results in poor muscle growth and excess fat gain. Likewise trying to refine preperformance psychological strategies without actual competition or performance demands may not achieve the desired mental state for competition.

Once all the major training phases have been outlined and the needs analysis has clearly defined goals over time, we can begin filling in the nutritional and psychological components of the annual plan. As seen in the following chapters, the preparatory, competitive, and transitional phases all have distinct training, nutritional, and psychological goals. These large-scale goals can be mapped into the existing annual plan and can be addressed in the same timescale hierarchy described above; begin with annual strategies and work progressively all the way down to session to session strategies. This may require collaborative effort within the sports medicine team; however clearly subdividing each phase with distinct goals based on the demands of training and competition provides a framework for the entire staff to work with.

Chapter 5

DEVELOPMENT OF MOTOR ABILITIES, NUTRITION, AND PSYCHOLOGICAL PLANS

Tudor Bompa, PhD / Boris Blumenstein, PhD / Iris Orbach, PhD / James Hoffmann, PhD

An athlete is capable of reaching peak performance only when one will benefit from the highest level of physiological development, particularly, when the motor abilities dominant in that sport will reach maximum capacity. In other words, nobody will be able to reach high performance before the dominant motor abilities will be at the highest level possible.

The ability to achieve the highest level of development of specific motor abilities is possible only if you follow a specific sequence. This sequence we call the *periodization of dominant abilities* in a specific sport. A brief description of each phase of development of strength, speed, and endurance is presented next.

PERIODIZATION OF STRENGTH

In most cases, strength is an umbrella term. In the reality of sports training, an athlete is not developing strength just to be strong! On the contrary, an athlete is trained to be

1. powerful, or to be capable of increasing the discharge rate of the muscles involved in the shortest period of time;
2. agile (capable to quickly change directions during a game or match); or to have
3. muscle endurance, the capacity to apply force against resistance over longer period of time, such as in cycling, aquatic sports, triathlon, or any other sports of long duration.

When an athlete is capable of displaying one of the these abilities, or a combination of them, at the highest level possible, colloquially we say one is in a *good shape*. The athlete has reached a great physiological capability and can play/compete with great potential. At this level of development, the athlete can reach maximum performance.

But one must also understand that this performance can be maximized only if the athlete has followed a periodized nutrition and psychological plan to address, in training, both the body and mind.

Periodization of strength has several variations, depending on the ultimate quality you want to develop. As such, these are two variations:

1. **Development of power/agility** follows these phases:
   Anatomical Adaptation (AA), Maximum Strength (MxS), Power (P) and Agility (A)
2. **Development of Muscle-Endurance** (M-E):
   AA, MxS, and M-E

A brief analysis of all these phases will allow us to explain some specifics of the methodology of training necessary to consider to achieve your athlete's goals (for further information regarding periodization of motor abilities, please refer to Bompa and Buzzichelli (2018).

1. DEVELOP POWER OR AGILITY

AA. Since the scope of training for AA is a progressive *adaptation* to strength training, the loads used are lower but progressively increased over several weeks: 30/40–50/60% of one repetition maximum (1RM). The duration of this phase depends on the background of the athlete: from a minimum of 3–6 weeks of training.

The main benefit of AA is to increase muscle's capacity to overcome specific loads and to progressively be ready to apply force against heavier loads during the *maximum strength* phase. But, equally important is to also *adapt the ligaments* of the main joints to progressively tolerate higher loads during the following training phases. In many situations, trainers tend to neglect the development of ligaments.

This training necessity is achieved only if the program is designed over several weeks, say 6, and by using low loads. Please remember that ligaments are as trainable as muscles are! Stronger ligaments are an insurance against injuries. And most injuries occur at ligaments and not at the muscle level!

MxS is fundamental to any sport requiring the development of speed, power, and agility. It is a prerequisite for the development of a powerful, fast, and agile athlete. The main scope of this phase is to train the ability of *recruiting* in action the highest number of fast twitch muscle fibers. Training loads, especially for athletes with a good background in strength training, become progressively much higher: from 60/70 to >90% of 1RM. Duration of this phase has to also consider the background of the athlete: often from a minimum of 3–6 weeks of training. Longer MxS phase than 6 is recommended for elite level athletes in sports such as football, throwing events in track and field, scrimmage in Rugby, sprinting events in track, bobsled, and baseball.

P and **A** are the direct beneficiary of increased MxS. If the level of MxS is inadequate (athletes who are not trained for 6 and do not handle loads of >90% of 1RM), gains in power and agility might be questionable. During this phase athletes are trained to apply force against resistance at the highest level possible (to *increase the discharge* *rate* of fast twitch muscle). Duration of this phase can be between 3 and 6 weeks. If the sports require high level of power and/or agility, these two abilities have to be maintained during the competitive phase.

Maximum attention should be also given to *agility* training. For far too long agility was considered to have evolved from speed. In other words, if one is fast, he or she was regarded to be agile as well. In reality, *agility evolves from strength*, from having strong leg extensors and flexor. This explains why sprinters look strong, are built strong, and, as a result, are fast. If you'll test some sprinters on an agility drill, it might surprise you how agile they are! *Strength makes one fast and agile!*

Sports that will benefit from the development of P/A are team sports, racquet sports, field and sprinting events in athletics, martial arts, boxing, wrestling, short distances events in swimming, ski jumping, and so forth.

2. DEVELOP MUSCLE-ENDURANCE (M-E)

A periodization model for M-E has to consider the following phases:

AA: Since most of the strength training is directed toward the development of M-E, the AA phase can be around 3–4 weeks. Same methodology can be applied as in the case of AA for power/agility.

MxS has the scope of increasing the difference between how much force is required during the competition in an endurance dominant sport and the athlete's capability while performing in an exercise in the gym. For instance, if a rower pulls during the race with a force of 60 kg (130 lb), the scope of MxS is to use seated arm pulls with higher force, say, 100 kg (220 lb). As a result of gains in strength in the gym, one can expect a positive transfer to pulling in the boat with higher force than before, say 80 kg (176 lb). As result of gains in MxS, the rower will increase the force applied against the water resistance and, as a result, increase the velocity of the boat during the race.

M-E is a training methodology that has the objective of performing many repetitions nonstop against a standard load. Taking the same example, a rower will perform during the M-E phase several sets against a standard resistance, say 4 sets of 60 repetitions with a load of 50 kg (120 lb). For more details about the M-E methodology, please see Bompa and Buzzichelli's (2018) book.

M-E can make the difference between low and elite level of athletes in sports such as aquatic sports, boxing, cycling, triathlon, long distance events in swimming, Nordic skiing, long distance events in speed skating, and any marathon type of sports.

PERIODIZATION OF SPEED

Anaerobic endurance should normally precede any specific speed training for most events and sports. The scope of anaerobic speed is to build a physiological base (glycolytic endurance) that will ensure the athlete to be capable to tolerate the fatigue encountered during maximum speed workouts that will be planned for the following training phases.

This type of training refers to repetitions of 400 m where the intensity will be medium, and as such, the athlete will be capable to perform several repetitions per workout (6×400–6/8×400 m, with a rest interval of 4–6 minutes). As the phase for maximum/specific speed approaches, the distance decreases, and the number of repetitions increases or stabilize at the desired number (e.g.. 6–10×300 or 8×200 m @ 75% of maximum speed on this distance).

Maximum speed starts with shorter repetitions (30–40 m). Distance will increase (say, to 50–60 m) only when the athlete will be capable to maintain a good form of running over the initial distance (40 m). If good form is not maintained over the selected distance, it means the athlete demonstrates an inadequate specific strength and specific endurance. Please note that even in a 100 m sprint, from 70 to 80 m on, athletes must have what is called *specific endurance* (speed endurance). Therefore, one has to increase the duration of maximum speed very carefully. A good/perfect mechanical form demonstrates whether the athlete has increase the duration of repetitions carefully and progressively.

Specific speed refers mostly to the type of speed performed in various sports, from football to baseball/basketball, and so on. In the case of sprinting in track and field or speed skating, specific speed also means to train the last part of the race, or speed endurance.

PERIODIZATION OF ENDURANCE

Most endurance-based sports must be relatively fast at the beginning of the race, to avoid falling far behind the leaders of the contest and maintaining a steady speed during the middle part of the race to be in a good tactical position for the finish. During the finish of the race/game/match, athletes must be ready and motivate themselves to spend the last drop of energy, with the highest psychological determination to finish in the race in the best position or score the most points to win a team sports contest. Such a tactical plan can be enhanced by following the periodization of endurance suggested below.

Aerobic endurance represents the fundamental quality for endurance-based sports. Via specific types of training (fartlek, interval training, short, medium and long repetition, etc.), athletes improve not only the aerobic qualities but also the ability to use energy efficiently in a steady manner. Steadiness in endurance-based sports actually means an even utilization of energy based on a functional steadiness of the main functions of the body.

Anaerobic endurance is essential for the start and poststart of the race/game but also contributes to a very good finish. The finish is often the part of the race/game/match that decides the final part of the race/match/game. Muscle-endurance (M-E) can clearly enhance the capacity of the athletes to apply force against resistance (gravity, water resistance, ground profile, opponents) during the start/poststart and during the finish of the race/game/match.

Specific endurance is referred to the ability of the athletes to adapt to the specifics of the race/game /match. Since the race/game is often unpredictable, athletes must rely on their endurance and their tactical capacities to react to the variations of the race/match/game. For several sports, specific endurance also means to be capable to use variations of rhythm of the game/race/match to surprise the opposition and score more points or apply the tactics selected by the coach.

Model training can be a very good method to develop aerobic and anaerobic endurance by selecting training methods that do enhance the start and poststart velocity but also the last segment of the race/game. Therefore, short and fast repetitions with higher velocity will enhance a successful strategy in the early part of the race.

Steady-state types of repetitions are beneficial for the mid-section of the race while repetitions with higher acceleration can enhance specific endurance for different elements of the race/game/match, including the finish. Maximum effectiveness for the last part of the race/match/game can be achieved when repetitions of higher velocity are planned to be executed under the conditions of fatigue. Fatigue the athletes first, and then plan several repetitions that will mimic the finish.

EXAMPLES OF ANNUAL PLANS: PERIODIZATION OF MOTOR ABILITIES FOR SELECTED SPORTS

Figures 5.1–5.11 illustrate integrated periodization for annual plans for selected sports. A brief analysis of an annual plan (figure 5.1) will better familiarize the reader with specific elements of the plan. In the top of the chart are listed the months of the annual plan, leading to the most important competitive phase of the year. In this example we refer to Competitive phase 2, from mid-May to mid-August.

However, in the sport of track and field, there are two major competitive phases: one for the indoor (mid-December to the end of March) and the second for the outdoors championships. Next we refer to the periodization of the dominant abilities in sprinting: strength and speed.

Please note how we have periodized these two motor abilities and what type of nutrition plans and psychological techniques we have proposed for each training phase. These techniques were selected to meet the needs of the athletes in all the phases of the periodization of the dominant motor abilities for sprinters. Equally important, nutrition and psychological techniques were selected in such a way that they must assist the athletes to cope with the psychological strain of training and to ensure the energy supply as it relates to a specific training phase.

DESCRIPTION OF DOMINANT ABILITIES FOR THE EXEMPLIFIED SPORTS

Each of our examples refers to a specific sport. For a more comprehensive understanding of the periodization process and the abilities necessary to target in each example, a point-form analysis will be made for each sport. As such we'll refer to the following:

Ergogenesis (in percentage) refers to the *dominant energy* systems of a sport, determinant for the athletes to function efficiently during competitions. Without knowing the dominant energy system in the chosen sport, the coach might not train it properly, and as a result, planned performance might not be achieved. This information will help the coach/ instructor to also determine the dominant motor abilities and how to periodize them. Often, in team sports this is called **time-motion analysis that** allows the coach to decide the duration of a rally, the pause between rallies, and whether the athletes are properly trained to tolerate the physiological stress of the game or parts of the game.

Limiting factor for performance refers to the weakest point in the athlete's motor arsenal, or the state of physical qualities in a given phase of the athlete's development. Knowing the weakest points of athleticism will assist the coach/instructor to concentrate on strengthening a quality that holds down the athlete from improvement, from reaching the real potential in competitions.

Training objectives refer to the natural conclusion of the above analysis; especially the source of energy and the state of the development of the dominant motor abilities will allow the coach/instructor to set the training objectives for the future. Any weakness from the past must become a *new training objective*. This is the main road to athletic improvement.

SPRINTING

Annual plan for sprinting is described below (see figure 5.1).

- **Ergogenesis:** 53% alactic (phosphagen), 44% lactic (glycolytic), and 3% aerobic (oxidative)
- **Limiting factors for performance**: reaction time, starting power, acceleration power, power endurance
- **Training objectives:** maximum strength, power endurance

NUTRITIONAL PREPARATION IN SPRINTING

Sprinters are typically very lean and muscular and greatly benefit from having high power to bodyweight ratio. Protein is paramount not only to maintain the great deal of muscle in the lower extremities but also to ensure adequate recovery from the inherently homeostatically disruptive style of training. Maintaining a low body fat and high level of muscularity is essential to sprint performance, so coaches and athletes should plan hypertrophy phases well away from major competitions to allow time to reduce body fat and achieve an optimal competition bodyweight. Sprinters may also benefit from supplementing with creatine monohydrate and beta alanine.

PSYCHOLOGICAL PREPARATION IN SPRINTING

A sprinter is a powerful, strong, and explosive athlete. In sprinting the reaction time and muscle relaxation constitute crucial factors for a successful run. The significant psychological skills/techniques should be self-confidence, biofeedback training, self-talk, imagery, self-regulation, concentration, muscle relaxation, and the optimal balance between the last two.

<table>
<tr><th>Month</th><th>Oct</th><th colspan="2">Nov</th><th>Dec</th><th>Jan</th><th>Feb</th><th colspan="3">Mar</th><th colspan="2">Apr</th><th colspan="2">May</th><th>Jun</th><th>Jul</th><th colspan="2">Aug</th><th>Sep</th></tr>
<tr><td>Training Phase</td><td colspan="3">Preparatory 1</td><td colspan="4">Competitive 1</td><td>T 1</td><td colspan="4">Prep. 2</td><td colspan="4">Competitive 2</td><td colspan="2">T 2</td></tr>
<tr><td>Strength</td><td colspan="2">AA</td><td colspan="2">MxS</td><td colspan="3">Power
Maintain MxS</td><td></td><td colspan="2">AA</td><td colspan="2">MxS</td><td colspan="4">Power
Maintain MxS</td><td colspan="2">Recovery/
Regener.</td></tr>
<tr><td>Speed</td><td colspan="3">Anaerobic
endurance
(tempo running)</td><td colspan="4">Maximum speed
Reaction time</td><td></td><td colspan="2">Tempo running
Maximum speed (short distances)</td><td colspan="6">Maximum speed
Speed – endurance
Reaction time</td><td colspan="2"></td></tr>
<tr><td>Nutrition</td><td colspan="2">BCA</td><td colspan="2">WS</td><td colspan="3">RP</td><td colspan="2">BCA</td><td colspan="3">WS</td><td colspan="4">RP</td><td colspan="2">AR</td></tr>
<tr><td>Psychological Preparation</td><td colspan="3">Learn and practice muscle relaxation, imagery, concentration, and self-talk techniques. Modify and apply the techniques in preperformance routine.</td><td colspan="4">Develop and apply preperformance routine; optimal balance between muscle relaxation and concentration; self-confidence; positive thinking, biofeedback training</td><td colspan="3">Muscle relaxation, music therapy, active rest</td><td colspan="3">Develop and practice the relevant psychological strategies</td><td colspan="3">Apply preperformance routine; optimal balance between muscle relaxation and concentration; self-confidence; positive thinking, biofeedback training</td><td colspan="2">Active rest; muscle relaxation</td></tr>
</table>

T = Transition; AA = Anatomical adaptation; MxS = Maximum strength; BCA = Body composition alteration; WS = Weight stabilization; RP = Recovery and performance; AR = Active rest; RR = Recovery/Regeneration

Figure 5.1 Integrated periodization for sprinting in track and field (2 peaks annual plan)

MIDDISTANCE EVENTS (800 M) OR OTHER SPORTS OF SIMILAR DURATION (1:30–3:00 MINUTES)

Annual plan for middistance events is described below (see Figure 5.2).

- **Ergogenesis:** alactic 10%, lactic 60%, and aerobic 30%
- **Limiting factors for performance:** anaerobic and aerobic endurance
- **Training objectives:** acceleration, anaerobic and aerobic endurance

NUTRITIONAL PREPARATION FOR MIDDISTANCE EVENTS

Middistance runners will generally be prioritizing carbohydrate consumption since their training is inherently glycolytic. Protein may be progressively reduced to the lower end of the recommended values to allow for a great amount of carbohydrate within the same calorie constrictions depending on how long the event is. Carbohydrate and fluid replacement following competition and training sessions should be emphasized.

PSYCHOLOGICAL PREPARATION IN MIDDISTANCE EVENTS

Middistance runners are required for high motivation and readiness for "extra" physical and mental effort during practice and competition. Therefore, the dominant psychological skills/techniques include regulation of the athlete's emotional level, goal setting, biofeedback training, self-talk, and self-confidence during the run. In addition, concentration should be on comfortable running pace, focusing on feelings and bodily functions such as muscle tension, heart rate, and breathing (association attentional strategy).

Month	Oct	Nov	Dec	Jan	Feb	Mar	Apr	May	Jun	Jul	Aug	Sep
Training Phase	Preparatory 1		Competitive 1			T 1	Prep. 2	Competitive 2			T 2	
Strength	AA		Power endurance			AA		Power endurance				
Speed			Maximum speed Reaction time									
Endurance	Aerobic endurance		Anaerobic specific endurance			Aerobic Endurance		Specific endurance			Aerobic	
Nutrition	BCA	WS	RP			AR	WS	RP			AR	
Psychological Preparation	Learn and practice muscle relaxation, imagery, concentration, self-talk, goal setting and disassociation techniques. Modify and apply the techniques in preper-formance routine		Develop and apply preperformance routine; readiness to extra effort during the run; motivational self-talk; optimal balance between muscle relaxation and concentration; self-confidence; positive thinking, biofeedback training		Active rest, muscle relaxation		Develop and practice the relevant psychological strategies; apply them in pre-performance routine		Apply pre-performance routine; optimal balance between muscle relaxation and concentration; self-confidence; positive thinking, biofeedback training		Active rest, muscle relaxation	

BCA = Body composition alteration; WS = Weight stabilization; T = Transition; RP = Recovery and performance
AR = Active rest; AA = Anatomical adaptation; MxS = Maximum strength

Figure 5.2 Integrated periodization for a middistance runner, 800 m (two peaks annual plan)

BASEBALL

Annual plan for baseball is described below (see figure 5.3).

- **Ergogenesis:** 95% alactic, 5% lactic
- **Limiting factors for performance:** throwing power, acceleration power, reactive power
- **Training objectives:** maximum strength, power

NUTRITIONAL PREPARATION FOR BASEBALL

Strength and power are key fitness characteristics for baseball players, so hypertrophy is often emphasized during noncompetitive periods. Baseball training is generally low volume with low work to rest ratios of activity, so baseball players may not require high doses of carbohydrate; however, they may require a substantial amount of total calories and protein to sustain body size and muscularity.

PSYCHOLOGICAL PREPARATION FOR BASEBALL

Successful baseball performance is determined by the player's ability to keep composure, confidence, attitude, and focus (Hanson, 2006). A high-level player must learn to maintain his focus on task-relevant cue (e.g., the ball for hitters and defenders, the target for pitchers) regardless of distracting circumstances. Therefore, the dominant psychological skills/techniques include relaxation, imagery, self-confidence, communication, self-talk, focusing, breathing, and routines.

<table>
<tr><th>Month</th><th>Dec</th><th>Jan</th><th>Feb</th><th>Mar</th><th>Apr</th><th>May</th><th>Jun</th><th>Jul</th><th>Aug</th><th>Sep</th><th>Oct</th><th>Nov</th></tr>
<tr><td>Training Phase</td><td colspan="3">Preparatory</td><td>Precomp</td><td colspan="6">Competitive</td><td colspan="2">T</td></tr>
<tr><td>Strength</td><td>AA</td><td>MxS</td><td colspan="2">Power
MxS</td><td colspan="6">Maintenance of reactive time
- Power
- Power endurance</td><td colspan="2">RR</td></tr>
<tr><td>Speed</td><td colspan="2">Anaerobic speed</td><td colspan="2">Max. speed
Anaerobic speed</td><td colspan="6">Maintenance of maximum speed</td><td colspan="2"></td></tr>
<tr><td>Nutrition</td><td colspan="2">BCA</td><td colspan="2">WS</td><td colspan="6">RP</td><td colspan="2">AR</td></tr>
<tr><td>Psychological Preparation</td><td colspan="2">Learn and practice relevant psychological strategies such as muscle relaxation, imagery, breathing, self-talk, focusing, self-confidence, and communication</td><td colspan="2">Develop pregame routine, shifting focus between external and internal perspective, blocking distracting external and internal circumstances</td><td colspan="6">Practice and apply pregame routine, maintain positive attitude toward games, communication, and self-confidence</td><td colspan="2">Active rest, evaluate personal and team achievements</td></tr>
</table>

T = Transition; RR = Recovery/regeneration; AA = Anatomical adaptation; MxS = Maximum strength; BCA = Body composition alteration; WS = Weight stabilization; RP = Recovery and performance; AR = Active rest

Figure 5.3 Integrated periodization for baseball (elite)

BASKETBALL

Annual plan for basketball is described below (see figure 5.4).

- **Ergogenesiss:** 30% alactic, 40% lactic, and 30% aerobic
- **Limiting factors for performance:** acceleration, take-off power, power endurance
- **Training objectives:** power, power endurance, maximum strength

NUTRITIONAL PREPARATION FOR BASKETBALL

Basketball typically has a long competitive season and numerous games per week, often increasing the risk for muscular atrophy. Basketball players should be conscious of their protein consumption so as not to potentially lose muscle during long competitive periods. Additionally nutrient timing practices will be emphasized as many training sessions and competitions will be in close proximity to each other, and glycogen repletion may become a limiting factor to subsequent exercise performance.

PSYCHOLOGICAL PREPARATION FOR BASKETBALL

Basketball is a team sport played by individuals; therefore it requires attention to both team and individual parameters. Basketball players usually are more receptive to especially four psychological skills, such as concentration, imagery, biofeedback training, and self-talk (Burke & Brown, 2003). In addition, team cohesion, communication, leadership, self-regulation, and motivation are required to maximize success in this team sport (Henschen & Cook, 2003; Lidor et al., 2007).

Month	Jul	Aug	Sep	Oct	Nov	Dec	Jan	Feb	Mar	Apr	May	Jun
Training Phase	Preparatory				Competitive							T
Strength	AA		MxS	Power Agility	Maintenance Power/agility							
Speed	An. end	Specific speed			Maintain specific speed							
Nutrition	BCA		WS		RP							AR
Psychological Preparation	Learn and practice relevant individual psychological strategies such as concentration, imagery, self-talk, and self-regulation; team psychological strategies such as communication, cohesion, leadership				Develop and apply pregame routine, preperformance routine (e.g., free throw, situations adaptation), team and individual confidence, self-regulation during game, attribution analysis after game; establish positive emotional team and individual atmosphere; biofeedback training						Active rest, muscle relaxation	

T = Transition; AA = Anatomical adaption; MxS = Maximum strength; RR = Recovery/regeneration; BCA = Body composition alteration WS = Weight stabilization; RP = Recovery performance; AR = Active rest

Figure 5.4 Integrated periodization for college basketball

FOOTBALL: LINEMEN (ELITE, COLLEGE)

Annual plan for football (lineman) is described below (see figure 5.5).

- **Ergogenesis:** 70% alactic, 30% lactic
- **Limiting factors for performance:** starting power, maximum strength
- **Training objectives:** maximum strength, hypertrophy, power

NUTRITIONAL PREPARATION FOR FOOTBALL (LINEMAN)

This is one of the few instances in sport where carrying extra bodyweight as ballast can potentially be very beneficial. Lineman are among the strongest and largest athletes on the planet, so they need to maintain high and consistent levels of calories at all times, particularly during hypertrophy phases and during the competitive season.

PSYCHOLOGICAL PREPARATION FOR FOOTBALL (LINEMEN)

American football is an interactive, continuous, contact/collision team sport. The linemen player is characterized by more strength than skill position, less running distance, and, on the other hand, maximum speed and fast reaction time. Psychological skills/techniques include goal setting, concentration, self-confidence, coping with pressure, communication, commitment, emotional control, self-talk, and awareness.

Month	Apr	May	Jun	Jul	Aug	Sep	Oct	Nov	Dec	Jan	Feb	Mar
Training Phase	Preparatory				Competitive							T
Strength	AA	Hyp.		MxS	Power Agile	Maintenance: MxS, P/A						RR
Speed		An. end		Max. speed React. time	Position-specific speed/reaction time							
Nutrition	BCA				WS	RP						AR
Psychological Preparation	Learn and practice relevant individual and team psychological skills such as goal setting, concentration, self-confidence, coping with pressure, communication, commitment, emotional control, self-talk, and awareness				Develop and apply preperformance and pregame routines, emotional regulation, attribution training, motivation, and self-confidence before and during tackling/during and after injury (when applicable)							Active rest, muscle and emotional relaxation

Hyp. = Hypertrophy training (body-building methods to increase muscle mass); BCA = Body composition alteration
WS = Weight stabilization; RP = Recovery and performance; AR = Active rest; RR = Recovery/regeneration

Figure 5.5 Integrated periodization for football: linemen (elite/college)

FOOTBALL: WIDE BACKS, TAIL BACKS (COLLEGE LEVEL)

Annual plan for football (wide and tail backs) is described below (see figure 5.6).

- **Ergogenesis:** 60% alactic, 30% lactic, and 10% aerobic
- **Limiting factors for performance:** acceleration power, reactive power, starting power
- **Training objectives:** power, maximum strength

NUTRITIONAL PREPARATION FOR FOOTBALL (WIDE BACKS, TAIL BACKS)

The backs positions perform a great deal of running and can sustain numerous contacts throughout games and practices. It is essential for them to replenish lost glycogen stores from strenuous games and training and maintain consistent protein intakes to ensure they are promoting recovery not only from substrate-level fatigue but also from structural damage.

PSYCHOLOGICAL PREPARATION FOR FOOTBALL (WIDE BACKS, TAIL BACKS)

The tail backs position requires quickness and agility as a runner as well as sure hands and good vision upfield as a receiver. Psychological skills/techniques include concentration and the ability of shifting attentional focus, anticipation reaction, imagery, muscle relaxation, self-talk, communication, and self-confidence.

Month	Apr	May	Jun	Jul	Aug	Sep	Oct	Nov	Dec	Jan	Feb	Mar
Training Phase	Preparatory				Competitive							T
Strength	AA	MxS	P/A, MxS	P/A	Power Agile	Maintain: P, A, RT						RR
Speed		An. end	Position-specific speed MxS, Acceleration Speed endurance			Maintain specific speed Max acc. Speed endurance						
Nutrition	BCA	WS			RP							AR
Psychological Preparation	Learn and practice relevant individual and team psychological skills/techniques such as concentration (attentional flexibility), imagery, self-talk, and muscle relaxation				Develop and apply preperformance and pregame routines, emotional regulation, self-confidence, imagery, self-talk and communication					Active rest, muscle and emotional relaxation		

T = Transition; AA = Anatomical adaptation; MxS = Maximum strength; P/A = Power and agility; RT = Reaction time; An. end = Anaerobic endurance; RR = Recovery/regeneration; BCA = Body composition alteration; WS = Weight stabilization; RP = Recovery and performance; AR = Active rest

Figure 5.6 Integrated periodization for football for wide backs and tailbacks (elite, college)

MARTIAL ARTS

Annual plan for martial arts is described below (see figure 5.7).

- **Ergogenesis:** 50% alactic, 30% lactic, 20% aerobic
- **Limiting factors for performance:** starting power, power endurance, reactive power, muscular endurance of short duration
- **Training objectives:** power, power endurance, maximum strength (<80%)

NUTRITIONAL PREPARATION FOR MARTIAL ARTS

Most arts are weight class sports and will benefit greatly from being as lean and as muscular as possible for that weight class. It's imperative that athletes are within ±2% of their target competition bodyweight at least one month out from competition, with one to three months being preferred to ensure optimal training and competing conditions. Hypercaloric phases to gain muscle mass should be placed early away from major competitions, and after some stabilization the martial artist can reduce the accumulated body fat with a hypocaloric phase to reach their target bodyweight for the major competitions of that competition cycle. However hypocaloric phases must be stabilized prior to competition; martial artists are notorious for cutting all the way into the competition, which can significantly alter their training and performance.

PSYCHOLOGICAL PREPARATION FOR MARTIAL ARTS

Martial arts events require quick response with high level of attention, self-control, consistency, and willpower during the match. Martial arts events share several specific characteristics and requirements of psychological skills/techniques such as psychological readiness, self-confidence, imagery, relaxation, biofeedback training, positive self-talk, self-motivation for creativity, and self-regulation to maintain an optimal level of concentration and anticipation (Anshel & Payne, 2006; Blumenstein et al., 2005).

Month	Jun	Jul	Aug	Sep	Oct	Nov	Dec	Jan	Feb	Mar	Apr	May
Training Phase	Preparatory							Competitive				T
Strength	AA	MxS	P/A RT	T; A/A	MxS	P/A RT/MT		Maintain: P/A; RT/MT				RR
Nutrition	BCA	WS		BCA	WS			RP				AR
Psychological Preparation	Learn and practice relevant psychological skills/techniques such as self-regulation, concentration, imagery, relaxation, self-confidence, and self-talk						Develop and apply preperformance routine before/between fights, self-confidence, concentration, and biofeedback training				Active rest, muscle relaxation	

T = Transition; AA = Anatomical adaptation; MxS = Maximum strength; P/A = Power/agility; RT = Reaction time; MT = Movement time; BCA = Body composition alteration; WS = Weight stabilization; RP = Recovery and performance; AR = Active rest; RR = Recovery/regeneration

Figure 5.7 Integrated periodization for martial arts (taekwondo, karate, judo)

ROWING AND KAYAK-CANOEING (1,000 M)

Annual plan for rowing and kayak/canoeing is described below (see Figure 5.8).

- **Ergogenesis:** 10% alactic, 15% lactic, 75% aerobic
- **Limiting factors for performance:** muscle-endurance, starting power, power endurance

NUTRITIONAL PREPARATION FOR ROWING AND KAYAK-CANOEING

Rowing sports are another instance where power to bodyweight ratio is very important to success. Another factor to consider is that some rowing sports such as sculling allow for the use of the legs, where others such as kayaking may not. This is relevant because in some cases increasing the size of lower extremity muscle mass may be beneficial, where in others it may be less beneficial or potentially detrimental. Coaches and athletes should be aware of maintaining a competition bodyweight that maximizes the canoers/kayakers' power to bodyweight ratio.

PSYCHOLOGICAL PREPARATION FOR ROWING AND KAYAK-CANOEING

Rowing and kayak-canoeing are characterized by high endurance and speed, an explosive start, and the maintenance of high speed and tempo throughout the entire distance of the race (Blumenstein & Lidor, 2004). Psychological skills/techniques include mental control, relaxation, mental toughness, focusing, biofeedback training, and mental self-regulation.

Month	Oct	Nov	Dec	Jan	Feb	Mar	Apr	May	Jun	Jul	Aug	Sep
Training Phase	Preparatory						Competitive					T
Strength	AA	MxS	T	AA	MxS	MxS M-E	Maintain; MxS, P-E					RR
Speed						Mx: speed	Maintain: Mx. speed					
Endurance	Aerobic endurance			Aerobic end. Aerobic end.			Maintain: An. endurance Aero. endurance					
Nutrition	BCA			WS			RP					AR
Psychological Preparation	Learn and practice basic psychological skills/techniques such as muscle relaxation, self-talk, imagery, self-confidence, high motivation, and goal setting			Develop and apply preperformance routine, concentration, awareness, self-talk, readiness for extra effort during race, biofeedback training, and self-confidence					Active rest and muscle relaxation			

T = Transition; AA = Anatomical adaptation; MxS = Maximum strength; M-E = Muscle endurance; P-E = Power endurance; BCA = Body composition alteration; WS = Weight stabilization; RP = Recovery and performance; AR = Active rest; RR =Recovery/regeneration

Figure 5.8 Integrated periodization for rowing and kayak–canoeing

SWIMMING: 100 M

Annual plan for swimming (100 m) is described below (see figure 5.9).

- **Ergogenesis**: 80% alactic, 15% lactic, 5% aerobic
- **Limiting factors for performance**: power, power endurance, muscular endurance
- **Training objectives**: power, power endurance, maximum strength

NUTRITIONAL PREPARATION FOR SWIMMING (100 M)

Short distance swimming follows many of the same guidelines of other strength/power sports such as replenishing lost glycogen and maintaining euhydration. Swimmers generally tend to have high training loads and greatly benefit from having a carbohydrate-rich diet. Short distance swimmers in particular may benefit from supplementing with creatine monohydrate due to the high-intensity short duration nature of the competition.

PSYCHOLOGICAL PREPARATION FOR SWIMMING (100 M)

Swimming for 100 m is an event in which the swimmer requires for monotonous practice with many sets, sustained speed, and short recoveries. During competition the swimmer requires for the ability to use maximum strength in a very short time. Dominant psychological skills/techniques include focusing, self-regulation of arousal, self-confidence, relaxation, biofeedback training, and thoughts and feeling control (e.g., Blumenstein & Orbach, 2012a,b).

Month	Sep	Oct	Nov	Dec	Jan	Feb	Mar	Apr	May	Jun	Jul	Aug
Training Phase	Preparatory 1				Competitive 1		T 1	Prep 2	Competitive 2			T 2
Strength	AA	MxS		P P-E	Maintain P/P-E		MxS	MxS P	Maintain P/P-E			RR
Speed	An. E	An. E accel	Mx speed starts; S-E		Mx speed S-E			Mx: speed	Mx speed, S-E An. E			
Nutrition	BCA	WS			RP		AR	WS	RP			AR
Psychological Preparation	Learn and practice basic psychological skills/techniques such as relaxation, imagery, concentration, self-confidence, self-talk, and coach-athlete communication				Develop and apply preperformance routine, concentration, imagery, relaxation, biofeedback training, and self-confidence		Active rest, music therapy, muscle relaxation	Develop and practice psychological strategies in training setting	Apply preperformance routine, concentration, relaxation, imagery, biofeedback training, and self-confidence			Active rest, muscle relaxation

T = Transition; RR = Recovery/regeneration; AA = Anatomical adaptation; P = Power; MxS = Maximum strength; P-E = Power endurance; An. E = Anaerobic endurance; S-E = Speed endurance; BCA = Body composition alteration; WS = Weight stabilization; RP = Recovery and performance; AR = Active rest

Figure 5.9 Integrated periodization for swimming (100 m, national class athletes, bicycle)

SWIMMING: 800–1,500 M (COLLEGE LEVEL)

Annual plan for swimming (800–1,500 m) is described below (see Figure 5.10).

- **Ergogenesis:** 10% lactic, 90% aerobic
- **Limiting factors for performance:** muscular endurance, aerobic endurance
- **Training objectives:** aerobic endurance, muscular endurance

NUTRITIONAL PREPARATION FOR SWIMMING (800-1,500 M)

Recommendations for longer distance swimmers parallel that of shorter distance swimmers, with the differences mainly coming from their training loads, which are inherently higher. Thus longer distance swimmers may have an additional carbohydrate requirement above that of their shorter distance counterparts. Longer distance swimmers may not reap the same benefit of creatine monohydrate supplementation during competition due to the longer nature of the events, though it still may benefit their training in other areas such as resistance training sessions.

PSYCHOLOGICAL PREPARATION FOR SWIMMING (800-1,500 M)

Swimming for long distance is an event in which the swimmer requires for level of endurance fitness over a period of time. Psychological skills/techniques include goal setting, muscle relaxation, concentration, imagery, motivation, establishing positive mental state, self-talk, biofeedback training, self-confidence, and coach-athlete relationship (Blumenstein & Orbach, 2012a,b).

Month	Sep	Oct	Nov	Dec	Jan	Feb	Mar	Apr	May	Jun	Jul	Aug
Training Phase	Preparatory				Competitive 1		T	Prep 2	Competitive 2			
Strength	AA	MxS	M-E		P-E M-E		MxS P-E	M-EL	P-E M-E		RR	
Endurance	Aerobic		Aerobic Anaerobic		Aerobic		Aerobic Anaerobic		Aerobic			
Nutrition	BCA		WS		RP		AR	WS	RP		AR	
Psychological Preparation	Learning and practice muscle relaxation, imagery, self-talk, concentration, goal setting, motivation, coach-athlete relationship				Develop and apply preperformance routine, positive mental attitude/ readiness, muscle relaxation, concentration, imagery, biofeedback training, self-confidence		Active rest, muscle relaxation	Practice and refine psychological strategies in training setting	Apply preperformance routine, muscle relaxation, concentration, self-talk, biofeedback training, self-confidence		Active rest, muscle relaxation	

T = Transition; RR = Recovery/regeneration; AA = Anatomical adaptation; P = Power; MxS = Maximum strength; P-E = Power endurance; An. E = Anaerobic endurance; S-E = Speed endurance; BCA = Body composition alteration; WS = Weight stabilization; RP = Recovery and performance; AR = Active rest

Figure 5.10 Integrated periodization for swimming (800–1,500 m, college level, bicycle)

VOLLEYBALL (COLLEGE LEVEL)

Annual plan for volleyball is described below (see figure 5.11).

- **Ergogenesis:** 70% alactic, 29% lactic, 10% aerobic
- **Limiting factors for performance:** reactive power, take-off power, power endurance
- **Training objectives:** power, reaction time, maximum strength

NUTRITIONAL PREPARATION FOR VOLLEYBALL

Competition day nutrition for volleyball can be challenging because typical matches are scored as a best of five, and under tournament conditions there may be multiple matches within the same day or two-day period. Fatigue will generally manifest through glycogen depletion and dehydration. Therefore maximizing glycogen repletion through proper nutrient timing and glycemic carbohydrates following games is paramount and can also assist in maintaining a euhydrated state. Volleyball players should bring sports drinks and glycemic snacks during matches and tournaments to quickly replenish lost carbohydrate stores and be prepared for subsequent games.

PSYCHOLOGICAL PREPARATION FOR VOLLEYBALL

Volleyball is a team sport that takes into consideration the size of the athlete, ability to jump, dig and set with finesse, and perform with agility. During the game, the player requires to transfer between six different positions. Psychological skills and strategies include shifting attentional focus, communication, relationship between players, leaderships, fast decision making, muscle relaxation, self-confidence, and imagery.

Month	Jun	Jul	Aug	Sep	Oct	Nov	Dec	Jan	Feb	Mar	Apr	May
Training Phase	Preparatory					Competitive						T
Strength	AA	MxS	P MxS	MxS	MxS P/P-E	Maintain: P P-E MxS						RR
Nutrition	BCA		WS			RP						AR
Psychological Preparation	Learn and practice psychological skills/strategies such as muscle relaxation, imagery, concentration, communication, and self-confidence					Develop and apply preperformance and pregame routines, shifting attentional focus, communication, and leadership						Active rest, muscle relaxation

T = Transition; AA = Anatomical adaptation; MxS = Maximum strength; P = Power; P-E = Power endurance;
BCA = Body composition alteration; WS = Weight stabilization; AR = Active rest; RR = Recovery/regeneration

Figure 5.11 Integrated periodization for volleyball (college level)

Chapter 6

PSYCHOLOGICAL AND NUTRITIONAL PLANS FOR EACH TRAINING PHASE

Tudor Bompa, PhD / Boris Blumenstein, PhD / Iris Orbach, PhD / James Hoffmann, PhD

While in chapter 5 we have referred to the methodology of how to apply in your training plans (figures 5.2–5.11) the concept of integrated periodization (IP), in chapter 6 we are discussing in details nutrition and psychological plans for each of the training phases of an annual plan. Therefore, the presentation (chapter 6) of the training phases of an annual plan should be viewed just as a review. Details regarding training methodology and planning are available in other books written by the authors of present book (please refer to References).

Considering the theory of planning and training methodology as the foundation, we'll illustrate how nutrition and psychological techniques are integrated together into a multifunctional concept, so important to produce the best athletes possible. In order to do so, we'll analyze several training phases with the corresponding suggestions regarding best nutrition and psychological techniques/plans. This is in fact the most important chapter of this book: how psychological and nutrition plans are integrated together with your physical, technical, and tactical strategies.

PSYCHOLOGICAL PLANS FOR EACH TRAINING PHASE: GENERAL CONSIDERATIONS

Psychological preparation involves two parallel processes: (a) psychological support for daily practice, which dominates the preparatory phase, and (b) psychological preparation for competition, which is dominant in the specific preparation and in competition phases. Psychological plans developed for these two processes involve psychological techniques/strategies and psychological skill development, while taking into consideration the athlete's personality and sport discipline characteristics. The above elements should be taken into consideration for integration with other training athlete's preparations (i.e., physical, technical, tactical). In this chapter the above will be described in detail.

In general preparation, athletes learn the basic structure and general versions of psychological techniques. For example, the progressive muscle relaxation techniques by Jacobson (1938) can be learned and practiced for basic understanding, without a time limit, in a laboratory setting. In specific preparation, the technique is modified according to a sport's demands and is practiced with a time limit (e.g., 1–2 min) in laboratory and training settings.

In the competitive phase, the psychological technique is combined with other psychological techniques into one intervention package as part of a preperformance routine. The psychological package should take into consideration the sport discipline, the athlete's personality, and competition conditions. This intervention package is applied in relatively minor competitions and is subsequently used in major competitions in precompetitive, preperformance, and postperformance routines. In the transition phase, the psychological technique is practiced separately from other techniques, while the main goal is active rest.

As this process continues, the skill of the athlete's self-regulation and relaxation is being formulated. In the general preparation phase, the athlete demonstrates the initial stage of relaxation (first stage). The skill is not yet stable and is practiced for a long time period (e.g., relaxing for 15 min). In the specific preparation phase, the relaxation skill is adapted to the sport's demands and the athlete begins to perform relaxation automatically (second stage). In the competitive phase, the athlete uses the relaxation skill together with other psychological skills and performs under competitive stress (third stage). In this phase, skills are being adapted to the sport's demands and the athlete's personality. Later the focus is on adjusting and adapting the relaxation skills to a variety of competition scenarios and stress distractions (fourth stage). The top world athletes successfully demonstrate their relaxation skills in stressful events, such as the Olympic Games and World Championships.

NUTRITIONAL PLANS FOR EACH TRAINING PHASE: GENERAL CONSIDERATIONS

Although many believe that nutrition interventions across an annual plan are dealt with independently as their own unique entity, they are inherently linked to the type of training and workloads being placed on the athlete. The training guides the nutritional demands, thus the need for integration over time. Luckily the systematic phasic approach to training provided by good periodization practices also provides a systematic structure for nutritional interventions as well. Table 6.1 summarizes the interrelation of training and nutritional goals across an annual plan (as also seen in chapter 2).

Table 6.1 Goal summary for both training and nutrition throughout an annual plan

Training Phase	Preparatory		Competitive	Transition
	General	Specific		
Training Goals	Developing sport and training skills	Mastery of training and sport skills	Alleviating fatigue and elevating preparedness	Link multiple competitive cycles together
	Elevate work capacity and general fitness	Elevate sport-specific strength, explosiveness, and conditioning	Continued improvement or maintenance of fitness	Alleviate physical and psychological fatigue
	Basic sport tactics	Team and individual tactics	Perfecting skills and tactics	Maintain basic fitness
Training Phase	Body Composition Alteration	Stabilization	Recovery and Performance	Active Rest
Nutritional Goals	Gain muscle or lose fat	Achieve and maintain within 2% of competition bodyweight	Shift macronutrient priorities to carbohydrate	Reduce stress of dieting
	Develop basic diet and hydration routines	Refinement of eating and hydration patterns	Maintain euhydration	Maintain a reasonable bodyweight
	Consistent monitoring of bodyweight	Fatigue alleviation	Maintain lean body mass	Maintain lean body mass

Initially nutritional interventions will be geared toward establishing fundamental habits for eating and hydration that will be continually refined and rehearsed over time. This will include elementary measures of monitoring bodyweight, daily calories, daily macronutrients, and hydration. For some these practices may be a challenge when starting a nutritional program, so it is imperative to establish these routines early on so that in subsequent cycles the athletes can not only refine these practices but then also spend more time trying to promote their actual goals like increasing lean body mass or enhancing recovery.

Later as the athlete grows accustomed to monitoring daily intakes and bodyweights, they can begin individualizing their diet strategies with nutrient timing and food composition practices tailored to their personal preference. Modifying bodyweight will become increasingly important to achieve an ideal competition body composition during competitive phases throughout the year. Finally competition-specific nutrition routines such as weigh-ins, single and multiple event competition nutrition, travel, and precompetition routines should be refined, tailored, and rehearsed to the athlete's individual preferences.

PREPARATORY PHASE

As the key phase in your arsenal to produce best athletes, preparatory phase incorporates several types of training, from strength, power, and agility to aerobic and anaerobic endurance. For each of these phases we'll present brief comments, the most important ones being reserved to nutrition and psychological protocols. These protocols are specifically related not only to the needs of the sport but, more importantly, to how nutrition and sport psychology can assist the athlete to achieve training objectives set for a given phase.

ANATOMICAL ADAPTATION

TRAINING OBJECTIVES

- Adapt the body (muscles and ligaments/tendons) and mind to build a solid foundation for the training phases to come
- Training loads: 30–60/70% of 1RM
- Number of reps: 8–15 (without experiencing anatomical stress)
- Number of sets: 2–>3
- Rest interval: 2–3 minutes

The **psychological** protocol for the anatomical adaptation (AA) phase is demonstrated in **swimming 100 m** (figure 5.10).

The AA phase is considered the basic foundation for the other training phases. Therefore, the psychological training in this phase is oriented toward the development of the following:

- Formulating a positive and optimal mood
- A high level of sport motivation
- Clear training goals
- Strong commitment and self-discipline
- Positive and open communication with the coach

To achieve the above, the athlete's psychological skills training (PST) program should include the following main psychological techniques:

- Self-talk
- Concentration
- Progressive muscle relaxation
- Goal setting

- Biofeedback training

From a psychological perspective, it is important to take into consideration the fact that the athlete has returned following the transition phase (active rest) and is beginning systematic training in the gym and swimming pool. Therefore, a special focus is placed on the athlete's mood and sport motivation. Muscle relaxation, goal setting, and sport motivation are important parts of PST program within the AA phase. For example, during the practice itself, the athlete concentrates on the quality of the movement, while using self-talk and concentration techniques to achieve best performance; during the rest interval (2–3 min) the athlete uses self-talk in order to maintain his/her motivation and focus on training goals. In addition, in the AA phase, the athlete learns and practices a basic version of muscle relaxation, which lasts for 10–15 min and can be used at the end of the training as part of the recovery. Biofeedback training can be an additional important means of improving the athlete's self-regulation and recovery.

Nutritional interventions for AA are primarily focused on providing sufficient energy levels to support training and promote recovery. Nutritional goals for AA include these:

- Establishing maintenance level calories
- Establishing daily protein needs based on the athlete's size and lean body mass
- Establishing daily carbohydrate needs based on the athlete's activity levels
- Establishing hydration routines before, during, and after training

Nutrition for AA is not as rigorous and demanding as nutrition for the development of hypertrophy or endurance; however since training for AA typically occurs early on in the training process, it is a perfect time to establish preliminary numbers and habits that can be used in the many months of harder training to come. This period also serves as a great time to further integrate nutritional and psychological practices to monitor the athlete's perception of effort, fatigue, and performance during training and make individual adjustments based on their feedback.

MAXIMUM STRENGTH (MXS)

TRAINING OBJECTIVES

- Adapt muscle to overcome heavy loads
- Increase the recruitment of fast twitch (FT) muscle fibers
- Training load: -concentric contraction: 70–95%
  -eccentric contraction: >120%
- Number of reps: 2/3–8/10
- Number of sets: 3–6

- Rest intervals: 2–5 minutes, depending on the load

The **psychological** protocol for the maximum strength (MxS) phase is demonstrated in **martial arts** (figure 5.8).

The objective of the MxS phase is to develop the highest possible level of force. Therefore, the psychological training is oriented toward the development of the following:

- Mental readiness and determination to make the maximum effort
- Strong commitment and self-discipline
- Intrinsic and extrinsic sport motivation
- Development of awareness and self-control regarding the athlete's mood

To achieve the above, the athlete's PST program should include the following main psychological techniques:

- Decision making and cognitive processes
- Instructional and motivational self-talk
- Fostering a competitive atmosphere among the athletes
- Concentration
- Progressive muscle relaxation and breathing
- Biofeedback training

From a psychological perspective, the most important issue is that the athlete should be ready for periods of hard training. The athlete is required in each training session to overcome heavy loads.

Therefore, on one hand, psychological support should focus on sport motivation and self-discipline and, on the other hand, on recovery and relaxation between sets. To achieve sport motivation and self-discipline, the athlete goes through decision-making processes regarding the reasons and the benefits of the MxS for success in this specific sport. In addition, a competitive atmosphere among athletes is important for challenging the athlete's motivation. During training, the athlete should use motivational self-talk (e.g., "Come on, I can do it, push hard...") between sets and instructional self-talk (e.g., comments relating to the techniques, such as "squeeze the muscle, proper breathing") during the sets. Moreover, the athlete should complete a self-report at the end of each training session using a 5-point Likert scale evaluating his/her level of mood, motivation, and energy. This will assist the athlete in improving his/her awareness of the overall training state. This self-report should be done on a daily and weekly basis.

To achieve recovery and relaxation between sets, the athlete needs to continue to practice breathing, muscle relaxation, and biofeedback training. These specific techniques should be used between sets, during rest intervals (i.e., 2–5 min), and at the end of training. It is important to note that at this stage the athlete will be able to apply the techniques in the basic and long version (i.e., initial stage of learning), while some athletes may be able to use the techniques in a short and modified version suitable to their sport demands. The athlete continues to practice with biofeedback training in order to improve his/her self-regulation and effective focusing.

Nutritional considerations for maximum strength development are not distinctly different from other resistance training outcomes; however because of the timeline in which maximal strength training occurs, there are some significant alterations to the nutritional program as a whole that are noteworthy. Maximal strength training generally occurs in the specific preparatory periods and has a reduced maximum recoverable volume (see Chapter 8) than training for AA, hypertrophy, or endurance, and thus the training volumes at this time are generally lower. The goals of nutrition during maximum strength training are these:

- Achieving and maintaining within ±2% of the optimal competition bodyweight
- Reducing the stress from dieting for body composition alterations
- Refining nutrient timing and food composition to match individual needs

Training for maximal strength will generally result in a reduction in the total training load from previous phases such as hypertrophy, AA, or other phases developing the athlete's work capacity. It is important to continually adjust daily calorie and macronutrient intake as changes in the training loads occur.

When the athlete has transitioned from the development of work capacity and/or muscle growth into the development of maximal strength, then the nutrition plan should move the athlete into maintenance level calories, which would be a reduction from previous weight gain or an increase from previous weight loss, and should seek to start progressively reducing the stressors of dieting. Here the athlete should have already established fundamental eating and hydration routines; they should no longer be facing the discomfort and stress of weight gain/loss, and refining how much they eat, when they eat it, and what foods they eat. This can come in many forms, such as choosing foods and times that don't lead to GI distress, bloating, frequent urination, rebound hypoglycemia, "sloshing" in the gut, nausea, or any potential distractors from performing well.

HYPERTROPHY TRAINING

TRAINING OBJECTIVES

- Increase muscle mass (i.e., linemen in football, heavyweight categories in wrestling, etc.)
- Training load: 40–70/80%
- Number of reps: 9–12 (to reach exhaustion at the end of each set)
- Number of sets: 2–5/6
- Rest intervals: 45–90 seconds

The **psychological** protocol for hypertrophy phase is demonstrated in **linemen in football** (Figure 5.6) as follows

The objective of the hypertrophy phase is to develop muscle thickness while achieving the highest number of repetitions possible in each set as well as achieving exhaustion in each set and in the total number of sets. The hypertrophy phase of training is appropriate for sports where large body mass is important (e.g., linemen in football, weight lifting). Therefore, not all athletes should go through this phase. The psychological support in this phase is oriented toward the development of the following:

- A high level of sport motivation
- Mental and physical recovery between sets and after completing the exercise
- Mental readiness for the extra effort

To achieve the above, the athlete's PST program should include the following main psychological techniques:

- Instructional and motivational self-talk
- Fostering a competitive atmosphere among the athletes
- Progressive muscle relaxation and breathing
- Biofeedback training

From a psychological perspective, the athlete should be ready to expend extra effort (i.e., exhaustion) and to achieve a quick recovery during rest intervals (i.e., 45–90 sec). This means that in this phase it is important that the athlete will be able to apply a short version of the progressive muscle relaxation. In addition, the athlete should go through a long relaxation session (i.e., 10–15 min) after each training session. At the end of the week, the athlete should practice relaxation while focusing on the positive points of the past training week (e.g., high motivation, using positive self-talk during training, good technique, achieving a weekly training goal). Special focus should be given to a competitive and positive atmosphere among the athletes in order to achieve successful training.

Nutritional goals for hypertrophy are slightly more distinct and rigorous than training for other outcomes. It is important to remember that training for hypertrophy is used not only for gaining lean body mass but also for maintaining lean body mass under hypocaloric conditions, which many tend to forget about. The same stimulus that causes the muscle to grow is ideal for the prevention of atrophy during unfavorable energy conditions. Therefore the nutritional goals for hypertrophy training are these:

- Gaining lean body mass (Hypercaloric) or maintaining lean body mass (hypocaloric)
- Determining calories needs to establish a surplus or deficit needed for muscle growth/maintenance (approximately ±250–1,000 kcal per day)
- Determining daily protein demands based on bodyweight and lean body mass
- Determining daily carbohydrate demands to meet the high training volume

Training for hypertrophy will generally result in the highest weight training volumes the athlete will experience for that training cycle. As a result calorie and carbohydrate demands will be elevated compared to AA or maximal strength training phases. Additionally measurements of bodyweight and body fat should be monitored to validate the athlete is gaining/losing weight at appropriate rates and gaining/ maintaining muscle mass. One cannot manage what they do not measure, so it is imperative during hypertrophy training phases that bodyweight and daily calories are monitored to ensure proper rates of bodyweight change are achieved.

POWER AND AGILITY TRAINING

TRAINING OBJECTIVES

- Increase the discharge rate of FT muscle fibers
- Use implements: medicine balls (2–6 kg), power balls (6–>20 kg) heavy implements, such as shots from track and field, other implements (throw cattle bells)
- Agility drills of varied difficulty/intensity and duration
- Number of reps: 6–10
- Number of sets: 2–6
- Rest intervals: 2–>3 minutes

The **psychological** protocol for power and agility (P/A) training is demonstrated in **sprinting** (Figure 5.2) as follows:

The objective of P/A training is to assist the athlete in generating high amounts of force in relatively short periods of performance time and to quickly change directions during a game or a match. The success of the P/A training is determined by the achievement in the MxS phase. In many sports, P/A should also be maintained during the competitive phase. The psychological support in this phase is oriented toward the development of the following:

- Attention focusing
- Mental readiness for fast reactions
- Anticipation
- High sport motivation

To achieve the above, the athlete's PST program should include the following main psychological techniques:

- Simulation training exercise program (STEP: Blumenstein & Orbach, 2012a, 2015)
- Response training program (RTP: Blumenstein et al., 2005)
- Progressive muscle relaxation
- Biofeedback training

In P/A training, athletes practice on power and agility as necessary abilities to be used during the upcoming competitive phase. From the psychological perspective, athletes begin to practice a combination of psychological techniques to be used in the competitive phase as part of a preperformance routine. For example, in the STEP, the combination of muscle relaxation and concentration is critical. These two techniques are extremely important for P/A abilities. During the STEP, athletes learn how to relax the muscles and concentrate on the act itself. Progress in STEP is ultimately linked with the athlete's ability to combine muscle relaxation and concentration (see more details on STEP in the competitive phase described in this chapter).

The main objective of the second psychological package, named RTP, is to improve the athlete's response to different situations. The computerized training program consists of three reaction time (RT) tasks: simple RT (1 stimulus, 1 response); two-choice RT (2 stimuli, 2 responses); and discrimination RT (2 stimuli, 1 response) (see further details on RTP in the competitive phase described in this chapter).

During practice on the STEP and RTP, athletes are exposed to gradually increased stress distractions. It is imperative to reemphasize the importance of the ability to quickly relax and concentrate. Biofeedback training strengthens the positive effect of the ability to transfer learning skills to training, all this in order to assist the athlete to continue and improve his/her P/A. The above protocol is suitable for different sports, such as martial sports, team sports, and tennis at Olympic and Paralympic levels.

Nutritional considerations for power and agility development generally follow the same guidelines as previously described for the development of maximal strength. The goals of nutrition during power and agility training are as follows:

- Achieving and maintaining within ±2% of the optimal competition bodyweight
- Reducing the stress from dieting for body composition alterations
- Refining nutrient timing and food composition to match individual needs

Just as previously described, during considerations for maximal strength development, nutrition for power and agility training first ensures that the nutrition plan should move the athlete into maintenance level calories, which would be a reduction from previous weight gain or an increase from previous weight loss, and should seek to start progressively reducing the stressors of dieting. The nature of power and agility training will require the athlete to make explosive movements such as accelerating/decelerating, changing direction, and rotating their bodies, all of which can become unpleasant if the GI tract is full of dense foods or liquids. Bloating and "sloshing" can make athletes feel sluggish, insecure, and unwilling to make maximal efforts due to the GI distress and awkwardness of being full of food or liquid. This is an ideal time for athletes to refine and tailor the timing, quantities, and food choices they make in the pre, intra, and post exercise times to their individual preferences to promote performance and minimize distress.

MUSCLE-ENDURANCE

TRAINING OBJECTIVES

- Increase the ability to apply force against resistance for a longer period of time
- Training loads: 30–60%
- Number of reps: 1–>4 minutes
- Number of sets: 2–>4 (duration of a set: 2–12 minutes, depending on the duration of a race)
- Rest intervals: 1–2 minutes

This information is relevant for endurance dominant sports, where strength is essential for the final performance. To be organized after the MxS phase, toward the end of the preparatory phase.

The **psychological** protocol for muscle-endurance (M-E) training is demonstrated in **long distance swimming** (figure 5.11) as follows:

The objective of M-E training is to assist the athlete to perform many repetitions against a moderate load, such as in long distance running and swimming. The psychological support in this phase is oriented toward the development of the following:

- Mental readiness for monotonous effort for a long time period
- Tolerating pain
- Keeping up concentration for a long time period
- High motivation

To achieve the above, the athlete's PST program should include the following main psychological techniques:

- Progressive muscle relaxation and breathing
- Concentration for a long time period
- Tolerance to physical and mental "pain"
- A functional music program
- Association/disassociation orientation
- Self-talk
- Biofeedback training

From a psychological perspective, the athlete should be ready to tolerate physical and mental pain during monotonous nonstop work, while maintaining a high level of technical performance. This means that at this phase it is important that the athlete will be able to concentrate for a long time period and to relax during the performance and between sets. To assist in achieving this, the athlete should know how to shift between disassociation and association orientations (e.g., shifting between external and internal stimuli, such as music vs. technique). The use of a functional music program should be based on the physical demands of the training and follow professional recommendations in this area (Karageorghis & Priest, 2012). Athletes use positive self-talk and biofeedback training for strengthening motivation and improving self-regulation and concentration.

Nutritional considerations for muscular endurance generally follow the same guidelines as previously described for the development of maximal strength. The goals of nutrition during muscular endurance training are as follows:

- Achieving and maintaining within ±2% of the optimal competition bodyweight
- Reducing the stress from dieting for body composition alterations
- Refining nutrient timing and food composition to match individual needs

Although most of the nutritional recommendations for muscular endurance will parallel what was previously described in maximal strength, however there will be some adjustments based on training volumes differences.

SPEED TRAINING

TRAINING OBJECTIVES

- Increase maximum velocity and the capacity to accelerate maximally, in linear or curvilian directions, as required by the selected sports
- Number of repetitions: for maximum speed: 6–>10
- Distance:
  -For maximum speed: 20–60 m

  -For speed endurance: 60–120 m
- Rest interval:
  -For maximum speed: 4–>5 minutes

  -For speed endurance: 3–4 minutes

The **psychological** protocol for speed training is demonstrated in **basketball** (figure 5.5) as follows.

The objective of speed training is to develop the athlete's ability to move efficiently and quickly. Speed is an integral part of every sport: football, basketball, sprinting in track and field, etc. Speed, like strength, filters down to other abilities and can significantly benefit all players. The psychological support in this phase is mainly oriented toward the development of the following:

- Maintaining an optimistic mood and high motivation
- Mental readiness to fast reactions
- Attention focusing
- Optimal balance between relaxation and concentration

To achieve the above, the athlete's PST program should include the following main psychological techniques:

- Progressive muscle relaxation and breathing
- Concentration
- Self-talk
- Imagery
- Simulation training exercise program (STEP)
- Self-regulation and biofeedback training

From a psychological perspective, players should be mentally ready to run fast with a good technical performance. Therefore, an optimistic mood, a positive and competitive team atmosphere, and focusing on the technical elements of the run are important factors for achieving the training goals of this phase. In addition, using self-talk (e.g., motivational, "I can do this," and instructional, "Focus on relaxing your arms") helps players maintain an optimistic mood and improve mental readiness for speed training. Moreover, the role of positive self-talk is to provide the player with a tool that helps him/her focus on what he or she wants to achieve in the present and overcome negative obstacle thoughts. Imagery is recommended as an effective tool to reinforce the neurological pathways associated with actual movement and can be used several times before performance (e.g., during warm-up, stretching, bench time, etc.). Self-regulation of the emotional state and awareness of the optimal balance between relaxation and concentration can be achieved by using the techniques of STEP and biofeedback training.

Nutritional considerations for speed development generally follow the same guidelines as previously described for the development of power and agility training. The goals of nutrition during power and agility training are as follows:

- Achieving and maintaining within ±2% of the optimal competition bodyweight
- Reducing the stress from dieting for body composition alterations
- Refining nutrient timing and food composition to match individual needs

In addition to considerations previously discussed for power and agility training, nutrition for speed may also consider supplementing their plan with creatine monohydrate, particularly if the training is being performed in intermittent bursts. Creatine monohydrate can potentially provide a slight boost to short duration high-intensity performances and is a widely accepted ergogenic aid. Although previous protocols suggested it must be initially "loaded" into the muscle in high doses, this is largely not true and can cause GI distress for some. Simply adding approximately 5 g per day (depending on the athlete's size) in a beverage of choice can elicit the same effects. Coaches and athletes should be aware that creatine will cause the athlete go retain more water and thus may need to adjust their bodyweight monitoring program to accommodate this difference.

AEROBIC /ANAEROBIC ENDURANCE

TRAINING OBJECTIVES

- Build a solid aerobic base that will help the athlete to withstand the strain of training and competitions
- Long duration steady state activities
- Duration of a repetition: -long duration interval training: 10 minutes–>1 hour
  -anaerobic endurance (second part of this phase): 1–>5 minutes
- Number of sets: -aerobic endurance: 2/3–8/10
  -anaerobic endurance (lactic acid tolerance): 1–3/6
- Sport-specific anaerobic activities (technical and tactical drills): 30–90/180 seconds
- Rest intervals:- aerobic activities: 1–2 minutes
  -anaerobic: 2–>3 minutes
- Aerobic compensation

The **psychological** protocol for aerobic/anaerobic endurance is demonstrated in **mid-distance running** (figure 5.3) as follows.

Training of aerobic and anaerobic endurance involves and requires special psychological support for the athletes. Every training session is considered to be a physical and psychological challenge for the athlete to overcome. It is important to train specific endurance, which is the flexibility of shifting from aerobic endurance to anaerobic endurance and vice versa, throughout the training and later on throughout the race/game/match. From a psychological perspective, the athlete practices the capability to endure conditions of physiological/psychological stress distractions as well as the ability to shift between aerobic/anaerobic endurance based on training demands. The psychological support in this phase is mainly oriented toward the development of the following:

- High motivation
- Mental readiness for pain tolerance and extra effort
- Clear training goals with an emphasis on the link with competitive goals
- Positive self-talk

- Muscle relaxation during the performance and rest intervals
- Coach-athlete communication and positive feedback

To achieve the above, the athlete's PST program includes the following main psychological techniques:

- Motivational and instructional self-talk
- Imagery
- Progressive muscle relaxation and breathing
- Fostering self-confidence
- Concentration
- Biofeedback training

From s psychological perspective, runners should be mentally ready in each training session to "win." Motivational and instructional self-talk, together with imagery and self-confidence, are psychological techniques that assist the athlete in creating a positive emotional background for dealing with the training demands. Muscle relaxation, breathing, and concentration are effective tools for successfully dealing with pain and being able to demonstrate extra effort during performance. Coach-athlete communication and positive feedback strongly impact on the athlete's ability to cope with the physiological and psychological demands of training. Finally, biofeedback training strengthens self-regulation skills for fast adaption to the aerobic/anaerobic training load within and between sessions.

Nutritional goals for endurance training primarily revolve around ensuring adequate daily and carbohydrate demands are being met, particularly in the pre, intra, and post training periods. Endurance training typically provides the highest total training volumes and work rates in all athletic endeavors; therefore providing sufficient energy for training as well as replenishing lost energy stores following training are paramount. The nutritional goals for endurance training are as follows:

- Dosing proper pre and intra workout carbohydrate to sustain the voluminous training while minimizing GI distress
- Developing glycogen repletion protocols following training bouts
- Maintenance of lean body mass
- Develop electrolyte and micronutrient support protocols to support perspiration

Due to the high workloads and work rates of endurance training, significant water and micronutrient loss from perspiration can become problematic. Developing hydration strategies throughout the day as well as following exercise, along with sports drinks and micronutrient support products, will be helpful in preventing dehydration and electrolyte imbalances.

Endurance athletes typically don't carry as much lean body mass as strength and power-based athletes; however maintenance of existing lean body mass can be problematic due to the inherently high catabolic nature of the activity. Carbohydrate dosing will generally represent the largest relative portion of total calories from macronutrients; however protein intake should generally not fall below about 1.3 g of protein per kilogram of bodyweight (or per kilogram of lean body mass) per day to ensure adequate recovery of muscle tissue.

COMPETITIVE PHASE

The competitive phase is the highlight, the time of the annual plan, when athletes are testing their progress in competitions against opponents. This is why the way a coach is approaching this primordial phase of athleticism is essential to facilitate the best result of the year. This imply not only the peaking strategy, the methods used to accomplish peak performance, but also the entire complexity of athletes' preparation: psychological, nutritional, and rehabilitation strategies.

TRAINING OBJECTIVES

Create the physiological, psychological, and nutritional environment to achieve maximum performance.

- Maintain sports-specific types of training employed during the latter part of the preparatory phase. Additional sport-specific training is an important necessity to achieve highest performance.
- Use taper strategies to elicit peak performance during main competitions.
- Permanently monitor fatigue and the factors that elicits it.
- Often rest intervals may exceed the standards, depending of the cumulative level of the athlete's fatigue.
- For sports where the duration of competitive phase is long, maintain maximum strength and power or muscle-endurance.

The **psychological** preparation in the competitive phase involves two parallel processes. One direction is the psychological support for continuous daily training (see chapters 5 and 6), and the second direction is the psychological preparation for competition (see chapters 3 and 6). In this section the focus is on the psychological preparation for competition in **martial sports** (figure 5.8).

During the competitive phase the major goal of the athlete is to find and practice the optimal psychological state for peak performance. At this stage, due to the competition schedule, the athlete in martial sports should be able to know how to reach his/her "peak" several times. This means that the athlete should know "what, how, and when" to use psychological tools for his/her best performance:

What: Awareness of the optimal emotional state for peak performance, taking into consideration the athlete's sport discipline and personality.

How: Practice the relevant psychological skills/technique combination under competitive stress distraction.

When: Bring everything together, physiologically and psychologically, at the target competition time through precompetitive, preperformance, and postperformance routines in different competition situations.

Psychological preparation during the competitive phase is demonstrated by the example of one of the best European judokas. During a 4-minute judo match, a judoka is required to exhibit a quick response with a high level of attention, self-control, consistency, and willpower. The competitive phase in our example took place during the months of April–May, while the main objective was achieving success in the European Championship in May. During this period the main goals were to strengthen the athlete's psychological skills and strategies specifically for this event. Therefore, the PST program in combat sports included muscle relaxation, imagery, the LMA approach, RTP, and the STEP approach, while being exposed to different levels of stressful factors.

The main objective of the RTP is to enhance the athletes' responses under real-life settings (e.g., combat sport contests). The program consists of several RT tasks, such as simple RT (1 stimulus, 1 response); two choice RT (2 stimuli, 2 response); and discrimination RT (2 stimuli, 1 response). A computer simulation is used, and several factors are adopted during training in order to expose the athlete to more real-life competitive situations. Among these factors are a video demonstration of actual combats, external distractions such as noise, and competitions between two athletes (e.g., judokas) performing the reaction-time tasks simultaneously (Blumenstein et al., 2005). General descriptions, sets of training, and procedures of RTP are discussed in detail elsewhere (Blumenstein et al., 2005; Blumenstein & Orbach, 2012a).

Following are accumulated data, levels, and demands that are required when applying RTP. Several *stress factors* accompanied RTP and LMA. The response tasks are performed under the following situations (Blumenstein & Orbach, 2012a):

(a) Ordinary laboratory settings

(b) Positive and negative verbal motivation (M+/M-)

(c) Performance under precise demands

(d) Reward/punishment for performance

(e) Performance under "true" competition noise (competition audio clips)

(f) Performance under "true" competition sights (competition video clips)

(g) A combination of situations (a)–(f)

In RTP special attention needs to be given to the ratio between "fast" and "slow" RT, beyond the importance of time reaction. To exemplify this point, based on the numerous mental sessions utilizing RTP in combat sport, we found that the time limit for "fast" RT was < 200 msec and for "slow" RT > 200 msec. This balance point can serve as a good indicator of a performance goal and can characterize performance quality. For example, a process was initiated with a 2–3/12–13 ratio (2–3 fast and 12–13 slow RT); following 1–2 months of training, the athlete achieved an 8/7 ratio (8 fast and 7 slow RT) or 10/5 ratio. Right before the competition, a 14/1 ratio or even 15/0 was recorded. We found three quality levels for combat sports, as can be seen in Table 6.2.

Table 6.2 Difficulty levels for RTP in elite combat sports (M±SD)

Difficulty Levels	Reaction-time (RT) tasks (ms)		
	15 Simple RT	**30 Choice RT**	**30 Discrimination RT**
Baseline-Initial	M=200–220±40–50 ratio 3–6/12–9	M=240–265±40–50 ratio 2–3/13–12	M=210–230±40–50 ratio 1–2/14–13
Level 1 *Low*	175–185±30–35 ratio 8–9/7–6	200–210±30–35 ratio 8–9/7–6	185–195±30–35 ratio 8–9/7–6
Level 2 *Medium*	155–165±20–25 ratio 11–12/4–3	175–185±20–25 ratio 10–11/5–4	165–175±20–25 ratio 11–12/4–3
Level 3 *High*	135–145±10–15 ratio 14–15/1–0	155–165±10–15 ratio 13–14/2–1	145–155±10–15 ratio 13–14/2–1

Table 6.3 shows the best results achieved between 1991 and 2010 by elite combat athletes before important and successful competitions before European and World championships and Olympic Games.

Table 6.3 Best results achieved by elite combat athletes before successful competitions

Sport Disciplines	Reaction-time (RT) tasks (ms)		
	15 Simple RT	30 Choice RT	30 Discrimination RT
Judo	107±15–20 ratio 15/0	136±20–25 ratio 14/1	117±15–20 ratio 15/0
Taekwondo	110±20–25 ratio 14/1	126±15–20 ratio 13/2	122±20–25 ratio 14/1
Fencing	105±25–30 ratio 13/2	117±20–25 ratio 13/2	97±25–30 ratio 14/1

The STEP is composed of special exercise and simulated competition situations alleged to improve the athlete's psychological skills and performance. This rationale is based on two main ideas that have been suggested by numerous researchers and practitioners (e.g., Blumenstein & Orbach, 2012a, 2018; Krane & Williams, 2015; Orlick, 1990; Weinberg & Gould, 2015). The first idea is that psychological skills associated with athletic success are trainable. The second idea implies that by simulating competitive situations in practice, athletes can learn to optimally transfer their training experience to competition (Moran, 1996). The exercises are practiced during mental training sessions, and the psychological techniques are used for positive transfer in order to improve performance during competition. For example, Moran (1996) suggested that recommended exercises for improving concentration can be the "concentration grid" (Harris & Harris, 1984) and the "pendulum" (Weinberg, 1988). These exercises are named "adversity training" (Loehr, 1986), "simulation training" (Orlick, 1990), "reaction-training program" (Blumenstein et al., 2005; Blumenstein & Orbach, 2012a), and "specific psychological program" (Blumenstein & Orbach, 2012a). In addition, popular psychological techniques that are used to improve athletes' attention skills and performance are, for example, goal setting (Weinberg & Gould, 2011) and preperformance routines (Moran, 1996).

To achieve progress in judo, the connection between *muscle relaxation* and *concentration* is important. The judoka in our example above reported that often during a match, his muscles were tense when he achieved high concentration, which interrupted his technique. It was important to teach the judoka how to regulate his emotions and muscles during high concentration. Therefore, a special exercise for this case—the "stop-reaction" exercise—was developed (Blumenstein & Orbach, 2012a). In this motor task the judoka held a wide stopwatch while his thumb pressed on the start-stop button. The main goal of this exercise was to start and stop the time as fast as possible. To achieve good results, the judoka was required to learn how to quickly and relax his muscles and concentrate accurately on the act. In the first meeting, the judoka's results were 0.13–0.14 seconds. After one month of training, the athlete's results in this exercise were 0.8–0.9 sec; a number of times the judoka performed the exercise in 0.6 sec. Moreover, accordingly to his reports, the judoka applied the learned skill of fast relaxation and concentration, within his daily practice and training matches. This exercise was used in many variations, where each time the judoka had to achieve different results based on his opponent's level. In one training session, 20–25 "stop-reaction" repetitions were performed. It is important to note that the data from the RTP, STEP, and LMA approach were a motivator for the judoka to continue his psychological skills training.

In this phase muscle relaxation and imagery, which were modified according to the sport and the training goals, were widely applied in lab and training settings. An example of a modified technique that was sport specific is the relaxation exercise the judoka performed for 1, 3, and 4 minutes with fast concentration for 10 and 20 sec, while being exposed to external stressors such as competitive noises, film fragments, physical contact, and specific demands. In addition, the judoka practiced an imagery scene of one of his fights for 4 minutes, the same time length as the actual match. During imagery the judoka gave a finger signal every time he made an attack. The number of attacks per minute was recorded and analyzed together with the judoka. An important aspect of the psychological training was the involvement of the coach during the meetings; specifically, the coach played an active role in some of the cases where the judoka practiced his mental readiness using imagery. The coach made comments to the judoka during the imagery, strengthening the vividness of the imagery scene by making technical-tactical comments against a concrete opponent, similar to a real-life match. As a by-product, this experience significantly improved the coach-judoka relationship, while increasing the professionalism of the coach and the psychologist.

The above approaches and exercises were geared toward developing precompetitive and preperformance routines to be successfully applied in competitions and to help the judoka achieve world-class recognition.

Major **nutritional** considerations for the competitive phase can be found in Chapter 2; however it is worth revisiting some of these goals as they can often be overlooked or misunderstood. The transition from the preparatory phase to competition phase marks a significant shift in priorities: instead of training for the enhancement of fitness adaptations, the priorities shift to enhancing performance and recovery to maximize the athlete's preparedness. Here they should virtually already be at or very close to their ideal competition bodyweight and body composition, understand their daily calorie and macronutrient needs, establish the timing and frequency of meals, the macronutrient composition of each meal, and have found foods that meet their goals without interfering with their ability to perform. The competitive phase is generally not a time to change any of these things, and they should be well rehearsed through previous trial and error and become routine for the athlete.

The only substantial change that occurs during this time is a large shift in macronutrient priorities to carbohydrate. Here fats can be reduced toward minimal values to free up more available calories from carbohydrate, of course while maintaining an isocaloric state. Protein will remain largely consistent with what was done in previous phases; however it can be reduced as well if energy levels and recoverability appear to be dropping. Reducing protein intake below about 1.3 g of protein per kilogram of bodyweight (or per kilogram of lean body mass) per day will be insufficient for maintenance of lean body mass; thus athletes and coaches should first consider reducing dietary fats and only reducing protein on a needs basis.

The competitive phase should also be used to refine any specific nutritional protocols for competition such as carb loading, weigh-ins, water manipulations, single or multi event competition day nutrition, travel, and others. Some of these processes require a little trial and error, so they should be practiced early, preferably during precompetitive times or for competitions that don't hold any major significance so that by the time the important competitions roll around, the athlete is already comfortable with the process. Most precompetition specific routines will be continually refined and improved over time, but athletes and coaches should avoid trying new things for major competitions. Stick with what you know, and experiment at a later time when less is on the line.

Maintenance of lean body mass and hydration often go without saying; however they should continue to be monitored throughout the competitive period. Too often coaches do not monitor body composition throughout competitive periods, and only after their post season testing, do they find out their athlete actually lost lean body mass throughout the season. This can be easily avoided if it is monitored over time. The competition phase is typically a low volume period of training, and if athletes end up training under maintenance values and are not monitoring their dietary intake, they can be at risk for losing muscle mass. Steps should also be taken to monitor changes in bodyweight due to perspiration from training and competition to ensure adequate hydration is maintained.

Chapter 7

LONG-TERM INTEGRATED PERIODIZATION FOR DIFFERENT ATHLETIC CLASSIFICATIONS

Tudor Bompa, PhD / Boris Blumenstein, PhD / Iris Orbach, PhD / James Hoffmann, PhD

From the age of early teens to that of a college or professional athlete, a successful training program has to consider a long-term progression. This long-term progression does not refer to only athletic training but also refer to the equally important areas of nutrition and psychological assistance/plans. This may mean that both children and parents should understand that high performance is not going to be achieved tomorrow or at the end of the season. On the contrary, it will take several years of diligent training to adapt, to adjust the body and mind, to the rigors of a well-organized and progressively planned training, before performance expectations may be achieved.

Any long-term training program has to actually be a *long-term integrated periodization* (IP), where the major considerations have to be not just technical and physical training but most importantly to integrate them together with nutrition and psychological education. Since an athlete has to be a complete being, this cannot be achieved without instructing the athlete in the area of nutrition and mental training.

As illustrated by tables 7.1–7.3 (at the end of the chapter), a long-term integrated periodization refers to three categories of athletic classifications, such as the school years (or junior athletes), college and professional athletes of sports training. In reality, this classification represents in itself a progression that begins with the initiation years in training (the junior-high and high-school age) to the highest level of athleticism, the college, and especially the professional level of athleticism.

If we are used to accept the needs of a progression in physical training, we should also recognize that the same long-term progression is necessary for the areas of sports psychology and nutrition plans.

The basic reference point of a psychological program is analyzing an athlete's development from a sport career perspective. In line with this, it is important to discuss a psychological protocol for long-term progress in different athletic classifications. Each of these athletic categories has its own characteristics, content to consider, and crises to deal with. The transition from one classification to another (i.e., from high school to college and then to professional sport) is often accompanied by crisis events in the athletic career and successful/unsuccessful attempts to overcome these crises. Research found that a crisis often can stimulate the athlete to grow and gain new experiences (Stambulova & Wylleman, 2014). This can be achieved by providing special psychological support during crisis situations in the athlete's sport life. In cases where attention is not paid to providing psychological support, it may lead the athlete wanting to retire from his or her sport. Therefore, a psychological plan should take into account the athlete's age, sex, and sport level.

The progression of nutritional programs over time is really a matter of continued refinement and personalization. Although this will vary between different athletes and different sports, the basic pattern is universal: start with simple fundamental concepts and refine with detail and complexity over time. In addition, athletes will find certain preferences through trial and error with food choices and timing issues, which can also be integrated once they have been established. So when young athletes are beginning to integrate nutrition into their training plans, it's imperative to not overburden with numbers and structure, as this can become complex and tedious even for highly motivated adults. Simple things like making sure they have some protein in every meal, eating before and after training, and hydrating regularly can go a long way in developing fundamental routine habits. As athletes move into collegiate sport, this can change to eating certain amounts of food at specific times relative to their training schedule and spending more time into optimizing their body composition. Finally, as they move into professional and adult sport, perfecting their daily, training, peaking, and competition-based routines down to exact numbers will become necessary. Adopting this level of complexity too early before fundamental habits have been established can become an unnecessary burden for the athlete.

For any given training stage and development of a specific ability (i.e., strength vs. endurance), the body has to supply an adequate fuel and a specific type of mental plan. This is in fact the scope of integrated periodization (IP). To design a long-term model, a complete physical, psychological, and nutritional program for athletes and coaches to follow.

HIGH-SCHOOL ATHLETES

For many children participation in sports start during the elementary school years or the early classes of junior high or high school. These years of development do require great attention not only for the content of physical training but also for the type of nutrition and psychological protocol young athletes are suggested to follow. During the school years, young people are going through very difficult stages of anatomical, physiological, and psychological development: prepuberty, puberty, and postpuberty. Therefore, there are major psychological issues that should be taken into account. First, the young athlete perceives him/herself as an adult, while in reality his/her social position is that of a youngster. It is important to understand that although athletes at this age feel they can successfully deal with the demands placed upon them, sometime this "false" feeling of independence can lead to a crisis by itself. This may be reflected in conflicts with parents, teachers, and coaches on their social status and role in society. This trend is significantly stronger among girls compare to boys, due to the girls' faster sexual development (Laemmle & Martin, 2013).

At this age there are many temptations outside the sport world, which may impact the athlete's sport motivation. The social environment at this age, especially friends and other interest outside of sport, can serve as a stress distraction to keeping up with a training regime.

The training at this age serves as the basis for the athlete's future development. In many cases, it is difficult for the athlete to understand the connection between hard training in the present and future athletic achievement.

The young athletes take part in many national competitions. The reaction to and evaluation of success and unsuccessful performance may have an effect on his/her sport motivation, adherence, and self-determination.

Dominant figures at this age are coaches, parents, close friends, and teachers—all of whom play an important role in the athlete's development at this age. Social support, especially from the coach, plays a major role in the athlete's stability and commitment to the training process.

Finally, depending on the sport discipline, the dominant trend for this athletic classification is the training process—more than the sport result. It is true that in some sport disciplines, such as gymnastics and swimming, it is possible to find world-class athletes competing in spite of their young age.

Due to the above issues, the dropout rate from professional sport at this age is relatively high across countries and sport disciplines compare to other ages. For example, in Australia, Belgium, and Germany, only between 17 and 33% of elite junior athletes make the successful transition to senior elite sports (Wylleman, Rosier, & De Knop, 2016). Therefore, awareness of this phenomenon and the appropriate psychological intervention are critical.

Nutrition for the high-school athlete must be made simple. As coaches, scientists, and athletes, we have the tendency to be control freaks, which in most cases is very positive. This is not one of those cases. Providing high school–aged students with a rigid sport nutrition plan to follow is akin to herding cats, taking a Wookie to a barber shop, or trying to calm a Labrador retriever puppy. In practice it's simply an exercise in futility as most athletes at that age lack the discipline, maturity, and frankly the motivation needed to carry out structured nutrition plans.

What can be effective is more of a simple inception of some ideas with some guidance. Here are some relatively simple and foundational habits that can be promoted at this stage:

- Encourage them to eat a little more than they were previously used to, or more than some of their peers might be eating.
- Make sure each meal has a basic protein source so they are not just eating soda and chips type meals.
- Encourage them to eat before and after training sessions.
- Encourage fluid consumption and hydration throughout the day and during exercise.
- Start eating less junk food.
- Begin simple monitoring of bodyweight with no major stress or added pressures.

Bodyweight and body composition can often be over emphasized at this stage, particularly for those that have a weight class or rely heavily on the athlete's power to bodyweight ratio. This is something that can be explored; however, it should not be the emphasis for the high-school athlete. High-school athletes are generally still physically maturing, continuing to make significant gains in strength, power, and endurance from their low training age, and may not be remotely close to establishing an ideal competition bodyweight just yet. So this can be a time for trial and error where the athlete is observing how they perform at different bodyweights; however the athlete should not be making focused efforts to gain or cut weight for their sport. Too often young athletes can be pressured into dysfunctional eating behaviors at a young as a result of sport. The pressures of maintaining a certain bodyweight or body composition should wait until later on in their development.

Often the type of training children are exposed to is often dictated by traditions or by the schedule of competitions in the selected sport. However, irrespective of age a child is exposed to training, one has to realize that training has to be:

- Well organized, where training days have to be alternated (especially for the elementary school children) with days of rest
- To follow a progressive stimulation, meaning that the load, the physical and mental challenge in training has to increase slowly, over several years, as per children's capabilities
- Build a good foundation before specialized training will become dominant. The better this foundation, the easier it will be for children to progress, to reach a higher performance, both mentally and physically
- To understand that for any growing child, physical training must also be duplicated by an adequate nutrition and psychological considerations and watchfulness
- Follow a *multilateral*, overall physical development. After all, sports training has to result not only in a certain level of performance but also in a healthy body and mind
- *Before a young person will become a good player, he or she has to be the best athlete.* That means to train one to be strong, fast, powerful, agile, and capable to endure the pain and hurt of athletic competitions
- For maximum physical and psychological benefits, children should also be exposed to a high variety of athletic and recreational activities
- Consider individual characteristics, both physical and psychological. Training doesn't have to be regimented and rigidly applied, but rather according to each individual's potentials
- Therefore, a long-term vision should expose children to a long-term progression: from elementary school years to high school, college and, eventually, for some, to professional sports

From the psychological perspective, there are some core principles that should be considered at this age.

Learn and practice: Young athletes should be ready to mentally cope with physical and mental challenges in competitions. They should maintain motivation and self-confidence and know how to cope with success and failure. The focus should be on learning, applying, testing, and refining their physical and mental skills/strategies, in both training and competition settings.

Coaches and parents should educate these young athletes about developing a positive attitude toward training and competition in order to maintain a positive sport-life balance. Finally, young athletes should learn basic ethical and moral issues specifically aimed at the athlete's sport discipline. For these purposes, various psychological skill/techniques can be used:

- Goal setting
- Relaxation
- Imagery
- Concentration
- Self-talk
- Biofeedback training (BFB)

ANATOMICAL ADAPTATION

The main scope of early years of training should be a progressive adaptation. All aspects of training, technical, tactical, and physical, should be planned by the coach or instructor in such a way that before moving to a new challenge, or training difficulty, the young athletes should have the time to adapt to the activity performed in the past few months. Since this type of adaptation refers to muscle and connective tissues (ligaments and tendons), we'll call it *anatomical adaptation*. Do not plan new training tasks before adaptation to the new load has been achieved.

From the physical training point of view, anatomical adaptation (AA) means these:

- A progression in strength training, starting with low loads, 30–40% of one repetition maximum (1RM), that over 2–3 years can be increased up to 60–70% of 1RM
- An increase in the velocity of maximum speed training from 20–30 yards/meter (y/m) at 60–80% of maximum velocity to 50–80 y/m with a maximum speed of 60–100%. Drills of technical or tactical training can also be increased, over time, from lower speed (during the learning stage of a drill) to maximum speed
- Build an aerobic base for sport where endurance is essential. Low intensity activity over a desired distance (2–3 miles/3–4 km) can be increased over time to longer distances and increased intensity

Psychological support in AA is oriented toward the development of the following:

- A high level of sport motivation
- Clear training goals
- A strong commitment and self-discipline
- Awareness of body/mind functions during training and performance

- Formulating positive and optimal attitudes toward the training process
- Open communication with the coach.

To achieve the above, the athlete's PST program should include the following psychological techniques:

- Goal setting
- Imagery
- Progressive muscle relaxation
- Concentration
- Instructional and motivational self-talk
- Biofeedback training.

In the AA phase, young athletes learn and practice a basic version of muscle relaxation (10–15 min), which can be used at the end of the training as part of recovery. Biofeedback (BFB) training is an effective tool for improving the athlete's self-regulation, which can help in formulating a positive and optimal attitude in training and competition settings. The main focus of this phase is on learning psychological strategies in order to keep up motivation and focus on training goals. Concentration, imagery, and goal setting are important parts of the psychological training (PST) program within this phase, which can help the athlete manage negative thoughts and feelings and eventually perform better. In addition to learning the psychological skills, it is important to discuss with the athlete when, where, and how to apply these techniques. At the end of each training week, the athlete should prepare a self-report that reflects on what was helpful and why so that the self-awareness and the motivation of the athlete will be strengthened. For example, the following can be used: imagery and self-talk in the warm-up/before performance/after practice/in pauses during training and the game, relaxation between attempts, and so on.

From the nutritional point of view, AA primarily revolves around increasing the athletes lean body mass. This includes not only the agonist skeletal muscle for their sport but also the connective tissues that support it, including tendons, ligaments, fascia, and bone. This is primarily achieved through increasing calories above the baseline eucaloric state and also through proper manipulation of macronutrients. Young and developing athletes who are relatively new to sport will likely already require a substantial increase in calories simply due to increases in activity levels and possible maturation effects in order to maintain a eucaloric state, and to generate the net protein synthesis and energy conditions needed for AA there will likely be an additional calorie requirement on top of that.

For non-youth-based programs, AA can also include losing excess fat mass in order to maintain an ideal bodyweight or body composition. The training stimuli to grow muscle mass can also serve to preserve muscle mass under hypocaloric conditions; thus training to gain muscle or maintain muscle under energy-constricted conditions is often virtually the same. AA phases can also be used to serve athletes who have spent time gaining a substantial amount of lean body mass and now require time to shed any additional fat mass that has accumulated as a result. This practice however generally should not be used for youth-based programs, as they are often still physically maturing and developing and often lack the maturity and grit needed for calorie-constricted diets. Implementing of AA in youth athletes will primarily involve increasing calories and increasing protein intake through wholesome foods.

INJURY PREVENTION

A good and progressive adaptation is extremely important for injury prevention. Often young athletes are exposed to injuries specifically because the instructor didn't take the time to allow the athletes to adapt to a specific physical task or for being exposed to high demanding competitions before an adequate adaptation has been reached. Often, poor technique of lifting and inappropriate technical instructions can also result in injuries.

Most injuries occur not at the muscle but rather at the ligaments level (a connective tissue that holds together the anatomical integrity of a joint). When a ligament is damaged by a mechanical stress, an athlete is exposed to injury. Therefore, a major objective of early years of training for young athletes is to prevent injuries. To strengthen ligaments and tendons to be able to withhold higher training and athletic stressors. This essential goal is achieved via a strength training program using low loads over longer period of time so that adaptation will occur (see the progression in Table 7.1).

Injuries are often experienced during prepubescence and pubescence age, when athletes are exposed to some high-intensity strength training activities such as Olympic weight lifting or power lifting. These athletes are often prone to injuries simply because they cannot activate their muscles as adults do (Dotan et al., 2012). Some injuries can be prevented if you would consider these words of caution:

- Do not abuse specificity! Constant exertion of same muscles, using same exercises, may lead to overuse injuries (Cain and Maffulli, 2005).
- Multilateral training, especially during prepubescence and pubescence, compensation activities for over-exposed joints, is a successful method for injury prevention.
- Do not participate in competitions in weight lifting, powerlifting, or use heavy loads, such as over 80% of 1RM, before athletes achieve skeletal maturation, often after the age of 18–20 years.

From a psychological perspective, there are some important points to consider in order to help in preventing injuries. First, the athlete's self-awareness should be high so that he or she will be able to read body/mind signals. Using self-regulation skills is helpful to achieve this. A major focus should be on quality training and adequate warm-up before practice and recovery after practice. The work atmosphere during the practice has to be professional, while being able to disregard distractions and maintain an attitude of strict self-discipline. Altogether, it is recommended that the coaching staff be aware of psychological factors such as stress and other life stressors, which may predispose the athletes to injuries, especially those who have poor coping skills (Weinberg & Gould, 2015).

Nutrition also plays a major role in the prevention and management of injuries. The food we eat directly stimulates the RA processes through blunting the catabolic effects of exercise, stimulating the anabolic recovery processes, but also provides the raw materials needed for tissue growth and repair. The mechanical trauma inflicted on muscle and connective tissues from training cannot circumvent the laws of conservation of mass or energy and will require a substantial amount of both in order to heal. Damaged muscle or connective tissues can be thought of similarly to an engine block in a car; if one of the engine components fail or is damaged, one cannot simply duct tape a replacement on top of the engine. It must be taken apart, and the damaged portions must be identified, removed, or possibly destroyed in order for the replacement to be installed. The muscle and connective tissues undergo a similar process, which requires a substantial portion of the consumed energy and raw materials. If those materials are not readily available, the body will breakdown existing tissue (other skeletal muscle) in order to accommodate the increase in protein turnover. So nutrition is an essential component of not only stimulating long-term fitness gains but also in promoting recovery and preventing injuries.

HIGH DEMAND TRAINING PROGRAMS

Following the first 2–3 years of adaptation, high-school/junior athletes can be progressively exposed to higher demanding/higher intensity training programs. This stage of athletic development, the late teens, has to be seen as a transition from junior athletes to college athletics: from moderate-medium intensity to higher intensity in strength, power, speed, and agility. In this way the last 2 years in high school has to prepare the high-school athletes for college-specific intensities.

The vast majority of athletic activities have to be seen as sport-specific, heavier loads (70->80%), maximum speed, power, and agility training. Specificity training, as requested by each sport, has to dominate most aspects of athletic training.

Most of the specifics of training for high-school athletes are illustrated by tables 7.1–7.3. Below please find a brief suggestion referring to the specifics of training for high-school athletes.

STRENGTH, POWER, AND AGILITY TRAINING

Most gains in strength and power are visible during the postpubescence stage (Behringer, Vom Heed, Yue, & Mester, 2010; Wild, Steele, & Munro, 2013), reflecting gains in testosterone and growth hormone (Hulthen et al., 2001). Increments in muscle mass parallel the development of sex organs (Rogol, Roemich, & Clark, 2002).

If strength training, mostly maximum strength using loads over 70–80% of 1RM, results in increasing the recruitment of fast twitch (FT) muscle fibers; power training, on the other hand, will stimulate the increase in the discharge rate of same muscle fibers (Enoka & Duchateau, 2008). Improvements in power are directly proportional to the level of improvement in strength. As such, when an athlete has to apply maximum force against medium loads, such as medicine balls, it will result in increased performance in throwing and jumping capabilities.

Agility, a highly regarded ability in most team and racquet sports, is not an independent athletic ability. On the contrary, improvement in agility is directly proportional to improvements in athletes' strength. Quick changes in the rhythm of running or changes of direction, such as acceleration-deceleration, are possible mostly as a result of improvements in strength. During deceleration leg muscles are storing elastic energy eccentrically, while during acceleration same energy is exerted concentrically.

From a psychological perspective, an important factor for developing strength, power, and agility is mental readiness for hard training. Athletes should be ready for intense workouts and overcoming heavy loads in each training session. During this period, self-discipline in the training sessions and outside requires special attention, including nutrition and rest. In line with this, psychological support should be focused on sport motivation, muscle relaxation, and self-discipline. During training the athlete should use motivational self-talk (e.g., "I can do it," "All the way," "Come on") throughout the training session. Special attention should be given between sets, during which instructional self-talk regarding the technical aspects of performance is suitable. In addition, the athlete can use muscle relaxation techniques and breathing exercises between repetitions and at the end of the training session. Biofeedback training is learned and practiced for better self-regulation and concentration. The above recommendations should be considered and adjusted based on the sport discipline.

From a nutritional standpoint on training for strength, power, and agility, the main factors generally revolve around adequate protein and calorie intake to support the growth and/or regeneration of the trained tissues. Recovery from this type of training can be suboptimal if sufficient daily calories are not met, which can lead to suboptimal gains in fitness or subsequent training. Athletes who are training for strength, power, and agility under hypocaloric conditions to specifically enhance body composition should just be more diligent in achieving all of the major dietary factors for success outlined in Chapter 2 to ensure adequate recovery from training.

SPEED TRAINING

Team sports coaches are constantly looking for fast athletes! Maximum speed is a highly regarded physical ability. To move fast during offense and defense is an important attribute of any athlete. However, what is often misunderstood is that speed directly depends on two elements:

1. Heredity, or the proportions between fast twitch and slow twitch muscle fibers (Bray et al., 2009). When this proportion is in favor of fast twitch fibers, such an athlete is said to have natural talent for speed
2. The quality of strength and power training an athlete is exposed to. Quickness and high speed directly depends on the ability of the muscles to contract forcefully, to apply force against bones, so that a limb can move fast. An athlete cannot be fast before being strong. Strength always precedes quickness!

From a psychological perspective, athletes should be mentally ready to produce fast reactions and movement by working on attention focusing and an optimal balance between muscle relaxation and concentration. Special attention should be given to the technical aspects of the athlete's performance. Therefore, using instructional self-talk (e.g., "relaxing arms," "where is the ball"), anticipation drills, fast decision making, and emotional self-regulation are effective tools for improving speed training. Biofeedback training can help young athletes better understand the optimal balance between muscle relaxation and emotional arousal. Muscle relaxation, imagery, and self-talk can help athletes improve their technical performance for speed training, which will lead to a better quality of speed training.

From a nutritional perspective, speed training does not necessarily require any major accommodations outside of possible supplementation with creatine monohydrate. The only other major consideration is that athletes should already be at or within ±2% of their competition bodyweight when training for speed, and not still making body composition alterations. Major changes in bodyweight and size may require a "recalibration" of technique and perceived effort. For example, sprinting after losing 10 kg of bodyweight may feel significantly different; likewise sprinting after gaining 10 kg of bodyweight will likely feel different. For better or worse this may require readjustment of the athlete's feel for their speed techniques, and bodyweight should already be stabilized near the competition bodyweight at this time.

ENDURANCE TRAINING

For sports of longer duration, endurance is the main quality necessary to succeed in the selected sport. However, only in long duration sports, such as marathon running, triathlon, road cycling, or long-distance Nordic skiing, energy is supplied via aerobic energy system. In most other sports there are specific proportions between anaerobic and aerobic endurance (please consider the proportions for energy delivery systems presented by Bompa & Haff, 2009).

For most sports endurance helps athletes withstand the strain and pain of training and competitions. Low level of endurance capacity results in faster accumulation of fatigue, which, in turn, affects the ability to concentrate on the athletic task, resulting in technical and tactical mistakes, such as accuracy of passing and shooting.

Improvements in endurance are visible during all stages of athletic development, culminating during maturation (Baxter-Jones & Maffulli, 2003).

For fitness enthusiasts moderate endurance activities have a positive benefit on health indices such as hypertension and heart rate (Iwasaki, Zhang, Zuckerman, & Levine, 2003). Cardiorespiratory diseases are rare for individuals involved in endurance activities (Prasad & Das, 2009).

From a psychological perspective, young athletes should be mentally ready for practice sessions that are monotonous and of long duration. The focus of most training sessions, depending on the training program, is on the athlete's readiness to "win himself/herself," meaning giving extra effort. To achieve this, a positive mental atmosphere in the group/team should be established, together with a positive attitude toward the training goals, as well as on learning how to stop and disregard negative thoughts and to focus on either technical or external elements. In addition, motivational self-talk is needed when the athlete is feeling tired. All these can be accomplished with the constant and strong guidance and support from the coach. To support endurance training, muscle relaxation, breathing exercises, and concentration techniques are effective tools for successfully coping with pain, raising the ability to demonstrate extra effort during performance, and making a commitment to continue training in the specific sport.

From a nutritional perspective, training for endurance does provide a shift in priorities from what was seen in training for strength, power, speed, and agility. Of course because of the large training loads associated with endurance training, calorie balance is an essential consideration; however within calorie balance there will be a shift in priorities away from fats and protein and toward carbohydrate consumption. Athletes who participate in endurance activities like marathon running, triathlon, and cycling typically are not as large and muscular as their strength sports counterparts, and accordingly they will require less protein. Additionally their volumes of training are substantially higher than what is typically seen in strength sports; thus their calories, and the proportion of calories from carbohydrates, will be substantially higher. Strength and power athletes who also require endurance training, such as Rugby, football (soccer), and other field sports may need to maintain protein intake while increasing carbohydrates at the expense of calories from fats to sustain their body mass and training loads.

COLLEGE ATHLETES

With the background acquired during the high-school/junior years, many athletes are joining the animated college athletics. During these challenging years, college athletes will decide whether they have the necessary talent and abilities to become a professional athlete or will join the armada of professionals working for various businesses.

From a psychological perspective, the young athletes should be mentally ready to transition to a higher academic level (from secondary to higher education); to higher standards in training and competition; and to greater commitment, including time and travel. Moreover, the young athlete has to cope with leaving his/her familiar environment, which includes family, friends, and club. Therefore, the athlete should be ready to adapt to a new psychosocial environment, which requires higher levels of independence, discipline, and responsibility. On top of this, the athlete should establish a new, high-level relationship and interaction with the coach, teammates, and professional staff. The expectation from the student-athletes is that they will be able to combine academic education with a sport career, which requires a higher level of self-regulation and the ability to prioritize and plan their time between training and education (Wylleman et al., 2016).

Nutritional considerations for the collegiate athlete will generally parallel that of the aforementioned high-school athlete, however now in much greater detail. Instead of saying, 'Be sure to eat protein in every meal', now the athlete must be eating a prescribed amount of protein in grams per day. Instead of eating and hydrating around training times, now each of their meals are specifically mapped out with macronutrient content by time of day. Instead of monitoring bodyweight and body composition, the athlete is seeking to strike the perfect balance of muscularity, movement economy, speed, and strength for their sport. This is done through periodic cycles of muscle gain and fat loss generally until they reach the specific preparatory phase, where their bodyweights becomes strategically stabilized for competition.

From the years of multilateral, anatomical adaptation and injury-prevention training, where a solid foundation supposed to been build, college athletes are exposed now to a world of *specificity* training. From this point on, except for the early preparatory and transition phase, all elements of training are very specific, where the scope of training is maximum adaptation to achieve highest performance possible.

Most of the specifics of training for college athletes are illustrated by tables 7.1–7.3. Below please find a brief suggestion referring to the specifics of training for college athletes.

STRENGTH TRAINING

In order to maximize athletic performance, strength training plays a substantial role. From team to racquet sports, track and field, water sports, skiing, and so on, most athletes will benefit from gains in strength. In order to maximize strength, you have to target the prime movers (the muscles that actually perform the technical moves). Except for compensation purposes, disregard any strength training exercises that are not sport specific. To use most of the Olympic weight lifting and power training lifts preached by some instructors will be a disservice to the needs of athletes who dream for high performance.

The aim of MxS is to increase the recruitment of highest number of fast twitch (FT) muscle fibers. The higher the number of FT involved in action, the higher the probability to successfully defeating the resistance that your athletes are exposed to (force of gravity, water, opponent, implement, etc.). Training loads, during the MxS phase, have to be between 70–>90%. To be efficient you have to use both concentric and eccentric methods.

From a psychological perspective, the student-athlete should have high motivation and be mentally ready for hard training work. Basic psychological techniques to be applied in strength training include relaxation, breathing, imagery, self-talk, concentration, and biofeedback. The student-athlete should widely apply psychological techniques that are more suitable to the real-life demands of the specific sport. This will be reflected in using and practicing the techniques in their short version under stress distractions. In addition, instead of using the psychological techniques separately, the student-athlete will learn how to combine the various psychological techniques as one intervention package. Examples might be the combination of relaxation and breathing, relaxation and imagery, self-talk and imagery, and so forth. Finally, the student-athlete should be able to widely use relaxation for the purpose of recovery at the end of the training.

One of the major nutritional considerations for strength, power, and agility considerations for the collegiate athlete is the achievement of an optimal competition bodyweight and body composition. This may seem obvious for weight class sports like wrestling and weight lifting, or sports where power to bodyweight ratio is critical such as gymnastics and mountain biking, but it is also important for all other sports as well. A Rugby player must not only be strong and powerful, but they also need to have the stamina to run for two 40-minute halves and potentially need the sheer ballast or weight to remain competitive in physical interactions like rucks and scrums. A marathon runner may find that being too small leaves them lacking the strength to manage hills, whereas being too heavy is simply too inefficient for the event duration. More deliberate planning should be made throughout the year to make appropriate bodyweight and body composition changes around competitive periods.

POWER AND AGILITY TRAINING

Both power and agility are directly dependent on the level of MxS. They are force-dependent. In other words they rely on the level of MxS. Low level of MxS will have a negative effect on both power and agility simply because the athlete won't be able to recruit in action higher number of fast twitch (FT) muscle fibers. As a result, the athlete's capability to apply force against resistance will be impeded. Same is true regarding the capability to quickly change directions during games.

Application of force against resistance (medicine balls, power balls, heavier implements) has to be as quickly as possible. In the early part of power phase, use a progressive increase of the weight of balls, from 2 kg (5 lb) to 6 kg (12 lb), or higher. After 2–3 weeks of progressive increase, stabilize the weight for few weeks, and 2–3 weeks prior to the beginning of the competitive phase, decrease the weight progressively down to 2 kg (5 lb). As a result of this progression in loading-unloading, power abilities of an athlete will be maximized

Physiologically, power training results in increasing the discharge rate of FT muscles, meaning that the athlete will be capable of displaying specific athletic actions explosively, with maximum quickness and rate of execution.

MxS has a similar benefits for the improvement of agility. During the acceleration-deceleration coupling, elastic energy is loaded and then unloaded. This means that during deceleration elastic energy is loaded eccentrically. The benefit of eccentric loading is visible during concentric actions, when the athlete employs same energy in a concentric manner. As a result, changes of directions will be performed swiftly and effectively without loss of time.

From a psychological perspective, student-athlete should be oriented toward the developing of the following: attention focusing, mental readiness for fast reaction, anticipation, and high sport motivation. To achieve the above, psychological support should include the following psychological techniques: short version of progressive muscle relaxation, response training program (RTP), simulation training exercise program (STEP), and biofeedback training (LMA version).

Nutritional considerations for power and agility training generally follow the guidelines previously outlined; however one major consideration that can also be made is tailoring hydration and food timing to minimize bloating or discomfort during training. Changes of direction and rapid accelerations may cause "sloshing" sensations, and carrying too much food in the stomach may lead to bloating, nausea, and the potential for vomiting.

SPEED TRAINING

Whether you refer to linear velocity or maximum quickness in tactical drills, the scope of college athletes is to constantly maximize everything that has to be performed swiftly in the selected sport. Therefore, the goals of most training sessions is to improve/maximize sports-specific speed and, as competitions approaches, to increase the capacity of the athletes to apply the gains in speed/ quickness during competition/ game.

However, just specific speed will not lead to maximize athletes' ability. Please remember: *one will never be fast before being strong and powerful!* Therefore, periodize your strength and power training in such a way that ultimately will lead to improvements in quickness and maximum velocity (Table 6.2).

From a psychological perspective, the psychological support will be oriented toward the development of mental readiness for fast reactions, optimal balance between relaxation and concentration, attention-focusing, and maintaining an optimistic mood and high motivation. To achieve the above, the student-athlete should use psychological techniques such as self-regulation, biofeedback training, STEP, and imagery, which focus on the technical elements of exercise. An important factor in speed training among student-athletes is achieving the training goals in the new, high-level training regime.

Nutrition for speed would also follow the same recommendations outlined above in power and agility training. Another potential consideration during speed training is that the volume of training is generally reduced substantially to allow the expression of speed; thus the calorie needs may also need to be adjusted accordingly. If the athlete enters a speed macrocycle eating the same calories they did during a period focused on AA, they will likely start gaining weight simply due to the reduction in training load and maintenance of calorie intake.

MUSCLE-ENDURANCE TRAINING

For sports where endurance is a determinant ability in the struggle to achieve high performance, such as water sports, Nordic skiing, triathlon, and road cycling, muscle-endurance (M-E) will be an essential physical ingredient. The methodology of M-E, as explained by Bompa and Buzzichelli (2015), represents one of the most innovative types of training necessary to achieve maximum performance in dominant endurance sports. Under normal conditions of applying M-E methodology, your athletes should not gain in muscle mass. If you'll notice that, you should decrease the load and increase the duration of a repetition.

Long duration repetitions against a standard load, usually between 30–50% of 1RM, with specified rate and rest interval, results in athletes' ability to apply force nonstop that ultimately will decrease the time during a race. For these sports decreasing the duration of a race actually means high performance improvement.

From a psychological perspective, the student-athlete should be oriented toward mental readiness for monotonous effort, high motivation, concentrating for a long period of time, tolerating pain, and muscle recovery. To achieve the above, the student-athlete's psychological support includes progressive muscle relaxation, concentration for a long time period, self-talk, and biofeedback training. Specifically to this student-athlete's qualifications and the training phase, it is important to produce an optimistic, positive atmosphere within the group and a good relationship with coach.

Nutritional considerations for endurance training will generally revolve around optimizing recovery strategies post training as well as periworkout nutrition and hydration strategies during their respective events. Because the training loads of endurance sport are generally very high, the collegiate athlete may have to rely more heavily on nutrient timing practices, as well as manipulating the GI of ingested carbohydrate to ensure that they fully replenish lost carbohydrate. Typical glycogen-repleting protocols post exercise generally recommend about 1 g of carbohydrate per kg of bodyweight per hour following exercise. For some this can be a tall order to consume that much carbohydrate alone, let alone any additional protein or fat sources. Athletes should begin experimenting with foods that can be consumed quickly and easy while fulfilling their macronutrient requirements and not causing excess GI distress.

Additionally may also need to begin experimenting with how they hydrate and eat during their events. Some sports like cycling may allow more flexibility in what foods one is able to consume during an event, whereas others like skiing or running may be more constricted. There is a massive variety of sports drinks, chews, goos, bars, and gels that can be implemented during an event, all of which may be tolerated differently by the individual depending on the actual event. Carry weight, GI discomfort, and energy content should all be assessed by the athlete to find ideal combinations on race day.

ENDURANCE TRAINING

Endurance-dominant sports have to employ training methods that addresses primarily to both aerobic and anaerobic endurance. Normal progression is from a solid aerobic base, nonspecific to sport-specific longer duration activity, to anaerobic and sport-specific endurance. During the competitive phase, aerobic, lower intensity endurance activity has to be used mostly for compensation, recovery-regeneration type of training.

Specific tactical drills can also be performed for the purpose of recovery from a day of high-intensity, high training demand. In such a case the scope of training is this: compensation and recovery. And not a taxing-type of session. Therefore, this type of drills has to be of longer duration/lower intensity/more relaxing-type of activity. During a training session, or part of it, where the scope is aerobic compensation, you may also focus on the manner in which specific techniques/skills are performed. However, always remember that aerobic compensation is your main objective.

From a psychological perspective, the student-athlete should be oriented toward a high level of sport motivation, the ability to concentrate for a long time period, tolerance for physical and mental pain, and mental readiness for a monotonous effort over a long time period. To achieve these, the psychological techniques that should be used are self-talk, association/disassociation orientation, concentration, relaxation, biofeedback training, and functional music. At the end of each training session and training week, the training of the student-athlete must include various recovery measures for injury prevention. In addition, the coaching staff should remember that the student-athlete needs to combine education and training activities.

PROFESSIONAL ATHLETES

For any athletic dreamer, professional sports represent the highlight of athleticism. It is the stage of supremacy in all the sports in the world. Many athletes would sacrifice everything just to excel in sports and become a professional athlete. And the motivation is not just money! But also the satisfaction of maximizing personal, peak performance.

To reach high level of athletic proficiency, athletes do not need just talent. Most importantly is to invest, to devote at least 10 years, often close to 20 years of dedication, discipline, determination, and hard work. For this type of athletes their work is their religion!

Training in professional sports is very challenging. There is a high rivalry and competition for specific position in a team sports. High levels of fatigue are always the undesirable result of these challenges. Therefore, athletes' time has be carefully planned so that there is also time for physical and mental recovery and regeneration after training and games/competitions. Consequently, the key words that describe planning athletes' training and free time are *efficiency and effectiveness*

From a psychological perspective, the athlete has become a full-time professional athlete. The professional athlete status requires a greater number of training sessions, an intense and rigid competition schedule, and an increased need to focus on recovery means. The major challenge of the professional athlete is to produce excellence and stability in his/her performance at the highest level. Therefore, the professional athlete must fully integrate PST within the training process, based on the periodization model. A consistent balance between body and mind is achieved by increasing the athlete's self-regulatory competency in training and competition settings. In addition, the athlete should automatically apply and perfect a preperformance and precompetitive routine that combines psychological techniques as one intervention package. This will lead to the athlete's psychological readiness to achieve peak performance in many competitions at a high level.

From a nutritional perspective, the professional athletes see the continued refinement of the fundamentals all the way down to specific individual preferences regarding food types, quantities, and timing, not only for training and competition but also for peaking and tapering periods, weigh-ins and water protocols, tournament preparation, travel, and all other related scenarios. Calories and macronutrients are planned for each macrocycle, and bodyweight generally will not deviate outside about 5% of the ideal competition bodyweight throughout the year. Food, supplement, and hydration preferences are well established to enhance performance, recovery, and minimize GI distress. Specific protocols like carb ups, water drops, rehydration, post weigh-ins, all should be well rehearsed and practiced so the athlete can feel confident leading into competition with no surprises or added stressors.

Most of the specifics of training for professional athletes are illustrated by tables 7.1–7.3. Below please fined a brief suggestion referring to the specifics of training for professional athletes.

STRENGTH TRAINING

Among the key elements of strength training is to increase the load in anything related to this essential ability. Properly use periodization of training loads from 70–>120% of 1RM. Employ both concentric and eccentric methods. In the latter case selected sports (throwing and jumping events in track and field, football, Alpine skiing, all the sports requiring quick changes of direction and high take-offs, etc.) will greatly benefit from eccentric training with high loads (often >120% of 1RM).

Exercises have to categorically address to *prime movers*, muscles that perform the dominant technical skills. Be very effective. Select the lowest number of exercises, number of repetitions and sets. Warm up, do your work, and get out of the gym! And see your psychologist and nutritionist!

Psychological support in strength training is mainly focused on maximum level of concentration to overcome heavy loads, recovery and relaxation techniques as part of the training process. This has unique importance regarding the length of the professional athlete's sport career, due to the heavy load and the increased risk of injury. Optimal results of recovery means can be achieved by the integration of psychological and nutritional programs for maximum strength training results. The professional athlete is usually well aware of his/her psychological package for sport motivation, self-discipline, and self-regulation within the strength training. Examples for psychological packages might be self-talk and concentration before each set; the coach's support and breathing techniques before exercise; and goal setting and focusing before each set.

Nutritional considerations for maximum strength development are not distinctly different from other resistance training outcomes; however because of the timeline in which maximal strength training occurs there are some significant alterations to the nutritional program as a whole that are noteworthy. Maximal strength training generally occurs in the specific preparatory periods and has a reduced maximum recoverable volume (see chapter 8) than training for AA, hypertrophy, or endurance, and thus the training volumes at this time are generally lower. The goals of nutrition during maximum strength training are as follows:

- Achieving and maintaining within ±2% of the optimal competition bodyweight
- Reducing the stress from dieting for body composition alterations
- Refining nutrient timing and food composition to match individual needs

Training for maximal strength will generally result in a reduction in the total training load from previous phases such as hypertrophy, AA, or other phases developing the athletes work capacity. It is important to continually adjust daily calorie and macronutrient intake as changes in the training loads occur.

When the athletes has transitioned from the development of work capacity and/or muscle growth into the development of maximal strength, then the nutrition plan should move the athlete into maintenance level calories, which would be a reduction from previous weight gain or an increase from previous weight loss, and should seek to start progressively reducing the stressors of dieting. Here the athlete should have already established fundamental eating and hydration routines; they should no longer be facing the discomfort and stress of weight gain/loss and refining how much they eat, when they eat it, and what foods they eat. This can come in many forms such as choosing foods and times that don't lead to GI distress, bloating, frequent urination, rebound hypoglycemia, "sloshing" in the gut, nausea, or any potential distractors from performing well.

The differences between the three athletic classifications for strength training are illustrated in Table 7.1.

Table 7.1 Suggested model for strength training, nutrition, and psychological plans for different athletic classifications

Athletic classification/ Type of activity	High school	College	Professional
Strength training, power, and agility	- Anatomical adaptation - Load progression (40–80% 1RM) - Power: medium/high intensity - Simple and complex agility drills	- Maximum strength (MxS): 70–90%, concentric-eccentric - Maximize specific power and agility - Muscle-endurance (M-E) low-medium-maximum loads	- Maximize each athlete's potential - Perfect MxS: concentric-eccentric (70–>120%) - M-E: medium-high loads - Perfect power and agility: medium-maximum loads
Nutrition plans	- Increase protein intake - Increase lean body mass - Develop fundamental eating and hydration routines - Begin basic monitoring of bodyweight	- Ensure requisite levels of daily calories and macronutrients are being met - Introduce nutrient timing practices - Optimize body composition for the competition bodyweight	- Optimize competition, weigh-in, and peaking routines - Perfecting daily routines with personalized/ particular food choices - Exact personalized timing strategies
Psychological plans	Learn and practice basic psychological techniques (i.e., long version) to be used during training and competition; main goal is educational orientation while focusing on each technique by itself	Develop and practice short psychological techniques to be used during training and competition; combine psychological techniques; psychological recovery after training	Perfect psychological packages, which include self-regulation techniques and recovery means; combine psychological and nutritional programs; positive self-talk and sport motivation

POWER AND AGILITY TRAINING

Power and agility training has to also be very efficient! To effectively challenge the athlete's neuromuscular system and, as a result, to increase power and agility. Maximize both of these essential abilities by using specific exercises. Don't waste time to do exercises you have done for the past 15 years, such as the ladder! There is no training benefit from this exercise. After so many years of repeating same exercise, your athletes have reached a plateau. No more benefits!

If you still believe that repeating something you have done for the past 15 years is still beneficial to your athletes, you are living in a world of...fantasy! Do you still expect any increase in the level of athleticism of your athletes by using Bossu balls? Are you delusional? Wake up! This is just a gimmick!

Instead, use more demanding agility drills. That challenges your athletes' neuromuscular system. Heat the ground more powerfully in some agility drills to get ground reaction, to increase power of your athletes' triple extensors. Use heavy vests to increase the load. Use heaviest power and medicine balls to increase the discharge rate of FT muscle fibers and, as a result to increase power.

Plyometric method can still be very effective. Use heavy vests to increase the load in some plyometric exercises, but prior to the beginning of the competitive phase, eliminate heavy vests. As a result, the discharge rate of the FT fibers will be maximized. To increase power you can use heavy implements, such as 20–30 kg/40–60 medicine balls (MB). For example, hold an MB in front of your chest, go into a half squat, followed by a jump squat, and throw the MB as far as possible.

Psychological support of the professional athlete as part of power and agility training includes psychological programs such as the RTP, STEP, and LMA, which are based on the use of technological devices. At the professional level, the athlete understands the importance of why and how to use the trained psychological techniques. The short version of psychological techniques that are being used at the professional level are provided under high-level stress distractions specific for the kind of sport and training period (e.g., time limit). The result of this can be seen in the ability of the professional athlete to produce powerful and fast reactions in a stable manner throughout a long time period.

Nutritional considerations for power and agility development generally follow the same guidelines as previously described for the development of maximal strength. The goals of nutrition during power and agility training are as follows:

- Achieving and maintaining within ±2% of the optimal competition bodyweight
- Reducing the stress from dieting for body composition alterations
- Refining nutrient timing and food composition to match individual needs

Just as previously described, during considerations for maximal strength development, nutrition for power and agility training first ensures that the nutrition plan should move the athlete into maintenance level calories, which would be a reduction from previous weight gain or an increase from previous weight loss, and should seek to start progressively reducing the stressors of dieting. The nature of power and agility training will require the athlete to make explosive movements such as accelerating/decelerating, changing direction, and rotating their bodies, all of which can become unpleasant if the GI tract is full of dense foods or liquids. Bloating and "sloshing" can make athletes feel sluggish, insecure, and unwilling to make maximal efforts due to the GI distress and awkwardness of being full of food or liquid. This is an ideal time for athletes to refine and tailor the timing, quantities, and food choices they make in the pre, intra, and post exercise times to their individual preferences to promote performance and minimize distress.

SPEED TRAINING

Once again be very selective with the methodology you use for the development of maximum speed. Try to constantly increase velocity, swiftness, and rapidity during tactical drills. Time some of the drills to measure the duration of specific, standard, tactical drills. Decreasing the time is a demonstration of improvement.

Continue to do strength, power, and agility training to have the necessary physiological support for these essential qualities in many sports. If strength and power decrease so does your athletes speed and agility! You cannot be fast if strength capabilities decreases!

Psychological support for the professional athlete in speed training mainly focuses on perfecting the behavioral pattern of the optimal balance between relaxation and concentration, developing fast reactions, and increasing attention-focusing. To achieve this, the professional athlete widely uses the short version of self-regulation with biofeedback training, imagery, self-talk, and STEP. At the professional level, the athlete is ready to use, as part of speed training, a portable Galvanic Skin Response Biofeedback device to perfect his/her self-regulation abilities (e.g., after warm-up, between attempts, at the end of training).

Nutritional considerations for speed development generally follow the same guidelines as previously described for the development of power and agility training. The goals of nutrition during power and agility training are as follows:

- Achieving and maintaining within ±2% of the optimal competition bodyweight
- Reducing the stress from dieting for body composition alterations
- Refining nutrient timing and food composition to match individual needs

In addition to considerations previously discussed for power and agility training, nutrition for speed may also consider supplementing their plan with creatine monohydrate, particularly if the training is being performed in intermittent bursts. Creatine monohydrate can potentially provide a slight boost to short duration high-intensity performances and is a widely accepted ergogenic aid. Although previous protocols suggested it must be initially "loaded" into the muscle in high doses, this is largely not true and can cause GI distress for some. Simply adding approximately 5 g per day (depending on the athlete's size) in a beverage of choice can elicit the same effects. Coaches and athletes should be aware that creatine will cause the athlete go retain more water and thus may need to adjust their bodyweight monitoring program to accommodate this difference.

The differences between the three athletic classifications for speed, power, and agility training are illustrated in table 7.2.

Table 7.2 An illustration of a model of speed, power and agility training, nutrition and psychological plans for the three classification of athletes

Athletic classification/ Type of activity	High school	College	Professional
Speed training	- Progressive increase of velocity and distance - Variety in intensity, power and agility using specific tactical drills - Use tactical drills of different duration and intensities	- Use technical and tactical drills with maximum speed and varied durations - Improve/perfect sport-specific speed - Maximize power and agility drills	- Maximize speed with sports-specific drills - Perfect sport-specific maximum speed - Perfect power and agility training - Reduce the time of standard tactical drills
Nutrition plans	- Begin estimating an ideal competition bodyweight or power to bodyweight ratio - Develop fundamental eating and hydration routines - Begin basic monitoring of bodyweight	- Optimize body composition for the competition bodyweight - Ensure requisite levels of daily calories and macronutrients are being met - Introduce nutrient timing practices	- Optimize competition and peaking routines - Perfecting daily routines with personalized/particular food choices - Exact personalized timing strategies
Psychological plans	Learn and practice basic instructional self-talk with a technical orientation; learn and practice biofeedback training for better awareness and self-regulation	Develop and practice optimal balance between muscle relaxation and concentration; attention to technical elements of student-athlete performance	Practice optimal balance between relaxation and concentration; using RTP, STEP, and LMA with high-level stress distractions; relaxation; imagery and self-talk

MUSCLE-ENDURANCE TRAINING

For any sport where endurance is a determinant ability, muscle-endurance (M-E) can be your secret weapon! Do not misjudge the benefits of M-E! If during the college years you have used low-medium loads for training M-E, professional athletes will improve performance visibly if the load will be increased to medium and high (40–50% of 1 RM).

If your athletes can perform sets of high repetitions with higher loads, you must expect real improvements. Let's take, as an example, 800–1,500 m swimming, rowing, canoeing, road cycling, water polo, or triathlon. What defines all these sports? Application of force against resistance for a long duration. The more force you can apply against water resistance and the force of gravity, the faster your athletes will cover the distance.

Yet, many of these athletes are still reluctant to do M-E. Maybe they didn't discover it yet! Maybe are still content with the traditions of their sport! Traditions are great. But sometimes can hold you back. Break the tradition, and do your M-E training during the latter part of the preparatory phase! You'll be surprise about its effectiveness in increasing your performance!

Psychological support during M-E training mainly focuses on concentration and relaxation for better coping with the monotonous work of this period. In addition, the coach's support and the positive group/team atmosphere are important factors in achieving good training goals. Special attention should be given to the psychological recovery itself and as combined with a nutritional program.

Nutritional considerations for muscular endurance generally follow the same guidelines as previously described for the development of maximal strength. The goals of nutrition during muscular endurance training are as follows:

- Achieving and maintaining within ±2% of the optimal competition bodyweight
- Reducing the stress from dieting for body composition alterations
- Refining nutrient timing and food composition to match individual needs

Although most of the nutritional recommendations for muscular endurance will parallel what was previously described in maximal strength, however there will be some adjustments based on training volumes differences.

ENDURANCE TRAINING

The effectiveness of your endurance training program depends on how sports-specific it is. Use sports-specific endurance training to simulate the proportions of anaerobic/aerobic training. After you increase aerobic capacity during the early part of the preparatory phase, the rest of this early phase of training has to be sport-specific dominant. As the competitive phase is approaching, implement more endurance training that has a higher proportion of anaerobic endurance.

Lactic acid–dominant training will become a necessity. On the foundation of a solid aerobic endurance during the beginning of the preparatory phase, in the second part of the same phase, you can challenge your athletes to adapt to higher proportion of highly demanding lactic acid training. This will allow your athletes to impose a higher intensity during the race/game and, eventually, to put the opposition in a state of distress.

Visible gains in endurance can become a successful strategy during games/races. Your athletes' high adaptation to a strong aerobic base can play an important role in the strategy you'll design for your them. From the second part of the preparatory phase, use anaerobic training methods to increase your athlete's potential to cope with the fatigue induced by a game played in a fast rhythm. Employ model training methodology (to mimic the specifics of a race/game) to enhance your game but to also prevent same negative outcomes that your team might be exposed to.

The main focus of psychological support for the professional athlete during endurance training is the ability to concentrate for a long period of time. To accomplish this, the athlete should be mentally ready for bearing pain and making an extra effort. The professional athlete is aware of when and how to use association/disassociation skills during training. Functional music is often a part of endurance training; the use of positive self-talk is essential. And coach support and a positive team/group atmosphere impact on the overall motivational level of the athlete.

Nutritional goals for endurance training primarily revolve around ensuring adequate daily and carbohydrate demands are being met, particularly in the pre, intra, and post training periods. Endurance training typically provides the highest total training volumes and work rates in all athletic endeavors; therefore providing sufficient energy for training as well as replenishing lost energy stores following training is paramount. The nutritional goals for endurance training are as follows:

- Dosing proper pre and intra workout carbohydrate to sustain the voluminous training while minimizing GI distress
- Developing glycogen repletion protocols following training bouts
- Maintenance of lean body mass
- Develop electrolyte and micronutrient support protocols to support perspiration

Due to the high workloads and work rates of endurance training significant water and micronutrient loss from perspiration can become problematic. Developing hydration strategies throughout the day as well as following exercise, along with sports drinks and micronutrient support products, will be helpful in preventing dehydration and electrolyte imbalances.

Endurance athletes typically don't carry as much lean body mass as strength and power based athletes; however maintenance of existing lean body mass can be problematic due to the inherently high catabolic nature of the activity. Carbohydrate dosing will generally represent the largest relative portion of total calories from macronutrients; however protein intake should generally not fall below about 1.3 g of protein per kilogram of bodyweight (or per kilogram of lean body mass) per day to ensure adequate recovery of muscle tissue.

The differences between the three athletic classifications for endurance training are illustrated in table 7.3.

Table 7.3 Suggested model endurance training, nutrition, and psychological plans for different athletic classifications

Athletic classification/ Type of activity	High school	College	Professional
Endurance	- Build aerobic base - Anaerobic and aerobic endurance of low, medium, and high demand - Sport-specific endurance	- Aerobic and anaerobic endurance of high demand - Sport-specific endurance - Compensation aerobic endurance	- Sport-specific endurance - Anaerobic endurance of high intensity - Compensation aerobic endurance
Nutrition plans	- Begin estimating an ideal competition bodyweight - Begin monitoring daily calorie intake and bodyweight - Develop fundamental eating and hydration routines	- Establish calorie demands to maintain ideal competition bodyweight - Develop glycogen repletion protocols - Develop periworkout/ competition hydration and eating strategies	- Optimize peaking and loading protocols for competition - Perfecting daily routines with personalized/ particular food choices - Exact personalized timing strategies
Psychological plans	Learn and practice muscle relaxation, breathing exercises, and concentration flexibility in order to be able to disregard negative thoughts and demonstrate extra effort	Develop and practice concentration during a long time period, to increase tolerance of pain, and as a means of recovery; shift from negative thoughts to positive ones and from internal to external attentional focusing	Automatic shift between association/disassociation orientation; positive self-talk; functional music; coach/ team support; recovery means integrated with nutrition

Chapter 8

FATIGUE, OVERREACHING, AND OVERTRAINING

Scott Howell, PhD / James Hoffmann, PhD / Boris Blumenstein, PhD / Iris Orbach, PhD

Most coaches and strength and conditioning professionals are familiar with the modified version of Han Selye's general adaptation syndrome as applied to sport training: alarm, resistance, and adaptation (Selye, 1984). In sport science, this concept evolved into the supercompensation cycle developed by Yakolev (Bompa & Haff, 2009). Several scientists have offered several ways to view the relationship between stress, fatigue, and adaptation within coherent frameworks. The concept of adaptive reserve and the fitness-fatigue model have been partially successful in meeting this end. However, no model has succeeded to merge current evidence on stress, fitness, fatigue, adaptation, and outside factors that covertly interfere with performance outcomes. Whereas the fitness-fatigue and current adaptive reserve models have effectively characterized the positive effects of adaptation and the interaction of fatigue, these only limitedly address the specific ways in which performance proceeds in directed adaptation.

Several elements of this broad puzzle are lacking and warrant closer attention. Currently, only acute stressors are addressed in an attempt to direct short-term adaptation and moderate fatigue over time. The transfer from acute stimuli to long-term adaptation is a large gray area. We know it happens, but the exact point and specifics are largely left to physiological conjecture and imagination. In response to this problem, we look to other scientific fields for answers. We find that some stressors do not align to the classical view of the adaptation response or even the accepted notion of stress. Indeed, many factors reside outside of what sports scientists normally address. The following is an attempt to broaden our view of adaptation, stress, and fatigue to include additional factors that possess the potential to affect sports performance.

ALLOSTATIC PARADIGM IN SPORTS TRAINING

We have all heard the term *homeostasis* and have a basic intuition of how this is applied in physiology. Homeostasis represents one model used to understand the fundamental regulation of physiological processes. However, there is another, less mentioned model termed *allostasis*. The former is the basis for our current understanding of the two major concepts used in sport science: the supercompensation curve and fitness-fatigue paradigm. The homeostasis model underlying these major ideas is very useful but has limited utility. To understand the difference between the two models, the basic characteristics of each need further explanation.

Sterling (2004) characterizes homeostasis as "stability through constancy" and allostasis as "stability through change." The differences in definition are obvious, but the meaning behind the definitions is less apparent. Consider the supercompensation curve, which describes physiological parameters such as the acute return to baseline. This is a direct reflection of homeostasis. A limitation is evident when we attempt to link this acute response into the long term. Here the term *adaptation* applies to modification of some fitness parameter. If we strictly subscribe a reference to baseline return, this is severely restricted and only offers a general heuristic to visualize a normal acute adaptive mechanism. This return to baseline achieves stability in a *constant* manner. On the other hand, allostasis recognizes that stability, in a physiological sense, is only realized through *change*; for example, chronic training continually shifts some physiological parameters. Here adaptation refers to the readjustment of response sensitivity to similar signals over time. Rematching of effect outputs to expected inputs is the central meaning behind this notion of adaptation. Allostasis in sport can be explained further by considering how we view performance. If our view regards the classical notion (homeostasis) as stabilization by constancy, then increased performance is clearly not *constant* or *normal* but characterized by *change*. The intent here is not to subdue either of these two models but present a broader and more realistic view of the adaptation process to align with current thought and observations from the field.

The classical model (homeostasis) addresses only singular higher control mechanisms (neural and somatic) and relationships between effectors in the scope of a *single training demand* or load. Allostasis considers increased performance as the cumulative result of multiple neural and somatic controls and the relationships of effectors that have adapted (changed) progressively to vigorous training given *chronic demand.* Increased training load from all sources causes physiological changes to multiple organ systems (set points) at any given time, which are dependent on the entire training process. In medicine, this increase in "allostatic" load leads to pathology, but in sport, this is required for the highest levels of performance. Further, systematic somatic and brain adaptations precipitate response patterns of "accustomed" or habitual sports performance. These "performance" patterns represent an "unnormal" state that is adjusted up and down, according to demand, and only realized with judicious integration of training methods leading to *chronic adaptation*. The training, nutritional, and psychological methods covered in this chapter are inextricably tied to these response patterns. Although the full extent of allostasis is well beyond the scope of this book, these areas are considered, by the authors, to be the primary factors involved in allostatic management, and thus optimal performance.

You may be questioning how another concept is relevant in a field with constant redefinitions. The answer can be illustrated by a medical analogy. For example, take an individual treated for a chronic condition such as hypertension. In almost all of those diagnosed with hypertension, the exact cause is unknown. However, we know there has to be an underlying cause. This can be explained by allostasis, but not homeostasis. In this example, the effectors still respond, in the notion of homeostasis, returning to baseline values, but this does not reveal why the patient's hypertension persists. The cause is explained in the allostasis model: the individual's blood pressure has adjusted upwards because the effectors have reset to meet a demand "perceived" by higher control mechanisms, thus establishing a higher set point. The same process is inherent to increases in performance.

In sports training, the allostatic paradigm can be thought of as a holistic process, in which multiple higher control mechanisms are controlled to allow the body to reset effectors to demand, thereby transferring acute responses into long-term adaptation to elevate performance. *Allostatic load* is a term used to explain physiological responses and consequences to all stressors, acute, chronic, or environmental on a given tissue, organ, organ system, or complete organism (Logan & Barksdlae, 2008). Allostatic management, relating to stimuli and fatigue, is a theoretical framework that considers all stressors perceived or otherwise and integrates training, nutritional, psychological methods to systematically influence high-level physiological controls to increase resilience and adaptive capability. The athlete with greater adaptive capacity (physiological resilience) has more sensitivity to increased demand imposed by vigorous training. Thus, the aim of allostatic management is to assess and control both positive and negative influences of stress from all sources to direct and facilitate the adaptive process. The scope of allostatic management goes beyond the classical stress response and fitness-fatigue paradigm to encompass any aspect that directly affects or has the inherent potential to increase allostatic load. Traditional interpretations of fatigue management generally focus on individual variables and how to modify them to balance homeostasis, whereas allostatic management considers system level homeostasis and focuses on multiple variables at any given point. This chapter will provide insight on allostatic paradigm through integration of physiology, nutrition, and psychology.

DEFINITIONS AND RELATIONSHIPS

Central to the allostatic paradigm are the concepts of fatigue, overreaching, and overtraining. Each concept exists on a continuum from fatigue to overtraining, characterized by accumulation of acute stressors, which increase the magnitude of allostatic load. Fatigue is defined as a complex phenomenon resulting from central nervous system, neuromuscular, endocrine, metabolic, and microtraumatic factors, which results in increased physiological responses and perceived effort along with decreased performance indicators (Taylor & Gandevia, 2008).

Fatigue is divided into two primary aspects: acute and chronic. Acute fatigue is movement specific, related to physiological mechanisms such as calcium cycling, cross-bridge cycling, altered pH, and diminished energy substrates. Chronic fatigue occurs when allostatic load increases beyond recovery capability. Physiologically, the ability to generate force is compromised, microtrauma becomes cumulative, energy substrates deplete, and neural excitation diminishes.

Overreaching is a consequence of short-term fatigue, planned or unplanned, that decreases performance temporarily until unloading or rest elevates performance beyond previous levels. Overtraining is operationally defined as failed allostatic management, resulting in a pathological condition, characterized by chronic illness and failure to adapt to training loads with resultant decreases in physical and mental capacity.

DIFFERENTIATION

Generally, the difference between allostatic concepts is the degree of severity and duration of stimuli. Fatigue is a component of overall stress that can be managed to allow short-term performance decrements to resolve over several days. Overreaching is an extension of short-term allostatic management that endures for up to two weeks. Each of the short-term components require unloading and/or rest for optimal recovery, unmasking of previous adaptations, and elevated performance. Overtraining, on the other hand, is characterized by excessive periods of performance decrements, maladaptation, and specific signs and symptoms indicative of a medical syndrome.

While acute systemic fatigue allows for increased adaptive potential via morphological and functional improvements, chronic systemic fatigue results in lowered adaptive response and overtraining. Overtraining involves metabolic, endocrine, and nervous system decrements, along with alterations in cellular signaling, energy substrate utilization, repair, and growth pathways.

CENTRAL AND PERIPHERAL FATIGUE

Fatigue is categorized into two distinct branches: peripheral and central fatigue. The former includes physiological consequences of the peripheral nervous system. Peripheral fatigue involves biochemical changes within functioning muscle that proceed to reduction in neural drive. Biochemical alteration is caused by metabolites such as hydrogen ions, inorganic phosphates, lactate, and calcium ions, among others. Each athlete has an "individual critical threshold" of peripheral fatigue related to afferent feedback and degree of sensory perception based on biochemical status, enzymatic function, task, and current preparedness (Amann, 2011). Central fatigue involves the brain and spinal cord, resulting in decreased desire, altered spinal cord synaptic input and transmission of upper and lower motor neurons in gray spinal matter (Taylor & Gandevia, 2008). The central nervous system loses the ability to excite motor neurons, leading to a loss of central motor drive and force-generating capacity. Given the difference between the two fatigue categories, central fatigue is not the primary day-to-day factor overemphasized by many specialists. Practically, the primary factor is peripheral fatigue.

SIGNS AND SYMPTOMS OF OVERTRAINING

Overtraining is a consistent process characterized by specific signs and symptoms and individual variability, dependent on several factors. There are physiological, psychological, performance, and perceptive measures used to assess overtraining. Symptoms are subjective evaluations that include inability to focus, irritation with daily tasks and people, decreased self-confidence, loss of appetite, and diminished inspiration for activities. Two major symptoms are disrupted sleep patterns and insomnia. Symptoms associated with overtraining can develop in response to altered neuroendocrine mechanisms, neurotransmitter production and function, and cumulative stimulation of the hypothalamic-pituitary-adrenal (HPA) axis.

Signs of overtraining relate to objective measures such as heart rate, reaction time, and hormone levels. Blood pressure has a tendency to increase in diastole into the clinical hypertensive range throughout overtraining. Heart rate increases at rest due to systemic increase in catecholamines produced by the stress response, altered ion channels located in cardiac tissue, compensatory mechanisms related to altered blood composition, hydration status, and altered vagal tone (Kumar, Abbas, & Fausto, 2005). A decrease in bodyweight may occur due to prolonged energy substrate imbalance, decreased tissue repair mechanisms, changes to the satiety center of the brain, and endocrine alterations. Other physiological consequences include decreased glycogen storage capacity, immune alterations, and depression of adaptive capacity.

One of the dominant mechanisms in the transition from short-term fatigue to overtraining involves the HPA axis and consequent hormonal and metabolic milieu. Chronic activation of the HPA axis, as a result of ineffective management of stressors, increases corticotropin-releasing hormone (CRH). This triggers the release of adrenocorticotropic hormone (ACTH) and subsequent release of cortisol from the adrenal cortex. Prolonged cortisol release is indicated in several pathologies and directly involved with overtraining sign and symptoms such as amotivation, depression, anxiety, weight loss, and decreased performance (Kumar, Abbas, & Fausto, 2005). In general, cortisol modulates glucose metabolism from nutrients, immune responses, and inflammation in both a positive and negative manner.

EFFECTS OF FATIGUE

Movement pattern and technique alterations are the first and most detrimental consequences of fatigue. As fatigue accumulates, speed, power, force-velocity expression, and force generation quickly decay. Reasoning, decision-making capacity, and speed of impromptu assessment diminishes. Research over the last decade has consistently demonstrated the negative effects of fatigue on athletes' concentration and accuracy of action. There is a general consensus that athletes will develop difficulties in focusing attention, concentrating, and executing decisions as fatigue builds toward the critical threshold. When fatigue increases, lapses in attention occur with increasing frequency, intensity, duration, and distraction. A study by Boksem and colleagues (2005) revealed performance deficiencies in fatigued subjects, characterized by lower reaction times, higher misses, and more false alarms with prolonged time on task. Interestingly, the subjects also had problems with involuntary shifting of attention to nonrelevant stimuli (Boksem, Meijman, & Lorist, 2005). This established that fatigue resulted in a reduction in goal-driven attention, leading to primary reliance on stimulus-driven attention.

All sporting actions are proportionately affected by degrees of fatigue. In sports that involve passing and shooting, accuracy and precision will decrease, resulting in more misses, less ability to generate force, and loss of attentional awareness. Sports involving sprinting and agility will observe a decrease in reaction time from a start or change in direction. Most sporting actions will become less efficient with shifting attention, increased perception of effort, and stimulus-driven attention.

NUTRITIONAL ASPECTS OF FATIGUE

One of the most powerful yet manageable contributors of fatigue comes in the form of substrate depletion—for example, the reduction of endogenous carbohydrate stores in the form of muscle and liver glycogen. Although fats and proteins contribute to energy production during exercise, carbohydrate is the primary energy substrate for virtually all sporting and exercise activities. It has been well established that athletes who do not replenish lost carbohydrate after intense or voluminous exercise are often subject to decreases in exercise and sport performance, decreased recoverability, hypoglycemia, decreases in cognitive abilities, and can even result in higher rates of protein degradation. Several studies have assessed the effects of various carbohydrate consumption protocols and the majority demonstrate that athletes who are not replenishing lost carbohydrate, taking in carbohydrate amounts matched to their activity levels, or just eating a relatively low amount of carbohydrate results in decreases in acute or chronic performance. Although there may be some merit in specific carbohydrate restricting protocols to enhance oxidative capacity in endurance athletes, which still generally results in decreased exercise performance, most athletes should seek to pair carbohydrate intake with daily activity levels to prevent substrate level fatigue.

In addition to substrate level fatigue from lack of carbohydrate, another major contributing factor to chronic fatigue is the accumulation of microtrauma in the soft and hard tissues. Overloading training is by definition inherently disruptive, and often results in cumulative trauma onto the muscles and connective tissues. A major component of managing this microtrauma is adequate protein consumption. Athletes generally need more dietary protein than non-athletes due to their increased muscle mass, increased metabolic demand, and the need to repair damaged tissue from intense and voluminous training. Although athletes may not focus on skeletal muscle hypertrophy at all times, the increased rates of protein turnover from exercise demands a higher and steady intake of protein to ensure adequate recovery.

Strategies to manage fatigue from a nutritional perspective generally revolve around replenishing lost glycogen, remaining hydrated or replenishing fluids, and taking in a sufficient amount of protein relative to the athlete's size and lean body mass. The effects of carbohydrate and fluid consumption, such as consuming a sports drink during or immediately after exercise, can also have immediate effects on performance and perception of fatigue during exercise, making it a popular choice during competitions and hard training sessions. Protein consumption, on the other hand, may not have the same immediate effects on the perception of fatigue or elevating blood glucose levels to maintain performance; however, it will ensure that protein turnover rates are favoring synthesis rather than degradation.

PSYCHOLOGICAL ASPECTS OF FATIGUE AND OVERTRAINING

The terms *fatigue*, *overtraining*, and *burnout* are widely discussed in the sport psychology literature. Coaches and sport medicine staff should better understand the psychological aspects and causes of fatigue, overtraining, and burnout to help prevent the occurrence of fatigue. In the sport psychology research and practical work, the main focus is on the terms *burnout* and *overtraining syndrome (OTS)*. The term *burnout* is defined as a psychological syndrome involving exhaustion (physical and emotional), depersonalization, sport devaluation, a feeling of low personal accomplishment, depression, low self-esteem, and finally a reduction in training efficiency (Hodge & Kentta, 2016). On the other hand, overtraining encompasses positive, short-term overtraining, called "functional overreaching," which may lead to positive effect as a result of concentrated loading. Negative training, labeled "nonfunctional overreaching," eventually leads to overtraining syndrome (OTS) in the absence of allostatic load reduction. Several factors such as training content, intensity and loads, recovery qualities, individual capacities and nontraining factors such as family and financial situation were found in many studies as key cases that may predispose athletes to OTS

(Goodger, Gorely, Lavallee, & Harwood, 2007). For example, athletes' perfectionism and motivational regulation offers an explanation of the possible relationship with athletes' burnout; balance between training load and recovery may be one of the major factors preventing OTS (Hodge & Kentta, 2016).

FATIGUE DETECTION METHODS

Up until the late 1990s, fatigue was assessed predominantly in clinical laboratories. Practitioners relied heavily on subjective indicators of fatigue. Currently, there are many inexpensive methods available in sport science. Heart rate monitoring is a practical detection method that records heart rate over the training cycle. The goal is to measure the heart rate upon rising every morning during intensive training and deloading. One of the first signs of fatigue accumulation will be increased pulse upon waking via greater sympathetic nervous stimulation. This fatigue indicator usually presents at 10–30 beats per minute over average baseline values for one week or longer. Grip strength is a reliable measure for assessing central nervous fatigue. Testing is performed unilaterally in the morning for each hand with a dynamometer. Loss of grip strength is a cost-effective indicator of neural drive suppression.

Another detection method is heart rate variability (HRV) monitoring. Researchers have established that higher HRV correlates to increased max (Achten & Jeukendrup, 2003). The use of HRV monitors can detect allostatic load during the training period by measuring the response to overall training load. This has proven to be a very effective fatigue detection device to regulate training loads. When HRV is low, extra active rest could be warranted. Conversely, when HRV is higher, more concentrated loading could be used. However, it should be noted that HRV is a lagging indicator, and the limitations should be interpreted as such.

Several self-report tools such as profile of mood states (POMS) and daily analysis of life's demands for athletes (DALDA) can be used to assess mood and behavioral changes that precede overtraining. Other less conventional tests involve biomarkers found in body fluids or blood. As fatigue accumulates, corticoid levels in the blood increase for several different reasons. Various saliva and blood tests can be used to measure cortisol levels that are readily available for purchase. The purpose is to establish baseline levels and make comparisons during different loading prescriptions. One obvious, but overlooked, indicator is performance decrements. This is especially true in team sports in which undivided attention to each athlete is not afforded. The careful eye of the coach should notice performance decrements that persist for two weeks or more.

Fatigue identification, alone, is worthless unless the data obtained is used in a comprehensive management system. Identifying measures to analyze over the training cycle of interest can be complicated, but also simplified into an effective surveillance protocol. The key is to pick at least three variables that can be reliably measured and teach the athlete how to monitor and record data. The practitioner should take as many measures as possible, but when necessary utilize the athlete. Information acquired by testing is only as good as the system in which it is used. Incorporation of periodization principles with testing protocols paints a more detailed picture of the training process and allows informed decisions that impact load selection, initiation of preemptive deload, or increasing intensity above programmed parameters. Allostatic management is practically realized through integration of physiology, nutrition, and psychology within a periodization model. No decision in the training process should be left to chance.

FATIGUE PREVENTION AND MONITORING

Proper planning, testing, and integration of periodization principles are the most important aspects of controlling fatigue and prevention of the transfer of nonfunctional overreaching into overtraining. The degree to which we plan for fatigue predicts athletic success. The overall aim is to prevent fatigue beyond a critical threshold. Although certain levels of fatigue are favorable for adaptation, management is vital to increase performance levels. Repeated and specific stimuli applied in training have certain cumulative physiological consequences that alters preparedness and increases allostatic load. Stressors outside training also contribute to fatigue and should be considered in athletic planning. Once training has ceased, fatigue must be ameliorated in a consistent manner.

Allostatic-based fatigue management contains several steps and considerations:

1. Periodization principles integrated at the physiological, nutritional, and psychological levels.
   - Limitation of overreaching in accordance with athletic ability and training age.
   - Frequent incorporation of unloading with decreased intensity and volume.
2. Training plans tailored to individual recuperative capacity (information taken from training logs and data measures).
3. Assessment of performance through measures specific to the athlete's sport.
   - Assessment of maximum recoverable volume (MRV).
   - Use of at least three measures: One performance, one physiological-based marker, and one psychological- or perceptive-based marker.
   - Objective measures are preferred over subjective measures.
   - Suggested markers include heart rate, HRV, illness, quantity and quality of sleep, perception of fatigue, rating of perceived exertion (RPE), feeling of helplessness, testosterone to cortisol ratio (T:C ratio), blood creatine kinase levels, performance decrements, and desire to train.
4. Use of testing data within a measurement system to gauge severity of fatigue and allostatic load.
   - Compare baseline and repeated measures to develop data specific to each athlete.
   - Utilize trend analysis to identify relationships over time.
5. Athlete instruction on how to self-monitor nutrition, daily stress, sleep, performance on the field, and differences in planned training periods.
6. Requirement of training logs with acceptable detail on training load, duration, bodyweight, rated perception of wellness, rated perception of sleep, injuries, illnesses, and so on.

7. Schedule of regular intervals to assess bloodwork, selected measures, and self-report data.
8. Utilization of effective management modalities.

MAXIMUM RECOVERABLE VOLUME

One of the most common questions regarding fatigue management is, "How much training should an athlete undertake?" Research on dose-response relationships has shown several major considerations when individualizing an athlete's training volume; however it does appear that the more training one can do, the more potential benefit. This however is confounded by the restrictions of the fitness-fatigue paradigm, which states that although training generates fitness, it also generates a proportionate amount of fatigue. Not training enough leads to poor gains in fitness, but training too much can accumulate detrimental levels of chronic fatigue, which left unabated can lead to nonfunctional overreaching and/or overtraining syndrome. Israetel and colleagues (2015) sought to address this issue of optimizing training volume through identifying a theoretical range of training and coined the term *maximum recoverable volume (MRV)*.

The MRV is a modern method to address allostatic load. When implementing MRV, the practitioner should consider what average values mean in the homeostasis model. The average value of any homeostatic mechanism does not suggest a set point but only the most frequent demand on the system. As fitness increases, the average value of the demand changes, and this requires reevaluation of the MRV. The constant change in fitness and fatigue requires a continual update of information. This is the practical reason why programming specific numbers for training loads is ineffective. In essence, the MRV considers both positive and negative consequences of adaptation and allows for reasonable prediction of training load in light of uncertainty. It is worth noting that virtually all physiological parameters (temperature, blood distribution, hormone levels) fluctuate with different amplitudes and time constraints with a singular aim: to realize the greatest physiological proficiency with least incurred costs.

The MRV represents the most training an athlete can perform from all sources while still being able to recover and continue making progress. It is the balance between the overload principle and the principle of allostatic management, ensuring the athlete is training as much as possible but still remaining within the capacity to recover and adapt. This range of training not only accounts for acute training tolerance at any given point in time but also accounts for cumulative changes resulting from short-term training adaptations throughout the training cycle. The MRV represents the range of training volume from the beginning of a training cycle, its accumulation, and the planned overreaching periods (if applicable). When evaluating the MRV, consider the following examples. For some sports such as powerlifting, weight lifting, and bodybuilding, this concept is fairly straightforward as most of the athlete's training comes from strength training and very little of anything else; thus evaluation is realized by the volume load of training. Other sports like mixed martial arts (MMA), decathlon, and CrossFit are much more challenging due to the sheer number of training activities that must be accounted for, with the additional challenge of dealing nonuniform units of measurement (volume loads, distances, duration, punches, shot throws, etc.).

When estimating an athlete's MRV, one of the first steps is identifying all of the potential training activities or stressors that could contribute to fatigue and allostatic load. The following is a noncomprehensive list of various factors that could be included in this estimation:

- Volume loads from strength training
- Distances or durations of conditioning activities
- Technical and tactical components of sport training
- Physical contacts (tackles, sparring, takedowns etc.)
- Flexibility and mobility work
- Speed and agility training
- Competitions

From the previous examples, if we look more closely at a sport like weight lifting, we observe a lot of strength training and some specific technique work, along with flexibility and mobility work. Thus, the majority of the athlete's MRV can be explained simply by their strength training volume loads. Moving up in complexity to field and team sports, we must factor in specific sport practice, strength training, conditioning, speed and agility training, and other factors. With the most complex sports such as MMA, we might include specific sport practices for multiple disciplines such as wrestling, Brazilian jiu jitsu, kickboxing, and many of the previously mentioned items like strength training, conditioning, physical contacts, and others. Estimating an athlete's MRV can be a potentially daunting task; however, there are many simple ways to develop a strong grasp on an athlete's tolerance to training.

First, in order to understand MRV, we have to have a better understanding of what we actually mean by recovery. Most biological systems define recovery as a return to baseline characteristics (homeostasis); however, for sport this leaves out a substantial piece of the puzzle, which is the gain in fitness or adaptation that we get from the training process (allostasis). When we discuss MRV, we are also considering the maximal adaptive volume, as one may be training hard enough to experience staleness but still able to recover and continue. Although this scenario is worth differentiating in science, for coaching we mainly consider the MRV because the difference between staleness and the ability to recover is too small to measure and the concept is frankly easier to understand. For sport when we consider what it means to be recovered, we are essentially referring to the ability to perform overloading training again. This may or may not be at the actual baseline level for fatigue or any given fitness characteristic but sufficient enough to generate overloading training.

Next, we need to monitor performance. This can be in combination with the existing testing and monitoring protocols for the team as well as any perceptive or psychological assessments. One of the quintessential markers of training above the MRV is performance consistently below expected overload or baseline levels. A strong monitoring program must be able to differentiate one or two bad training days from the effects of accumulated fatigue. In this case, training the above the MRV would likely manifest poor performance in several subsequent sessions. In the weight room, this is relatively easy to observe. Athletes who could normally squat a weight for 10 reps now can only get 6 or 7. Weights that were moderately challenging now appear to be extremely difficult. Technique and bar speed will deteriorate, and the athlete's maximal and submaximal efforts will be diminished. Similar effects can also be seen in field performance and all other training components. Obviously, the goal of achieving the MRV is not to necessarily detect early effects of overreaching but to rather estimate by trial and error the upper limits of the athlete's training tolerance. The intent is to outline the boundary of volume to ascertain how much is too much and how much is realistic.

When manipulating training volumes from all sources, it is important to increase the volume slowly and progressively over time while continually monitoring the athlete. Sets and efforts can be increased week to week gradually until warning signs or accumulated fatigue can be observed. For example, we might start a training cycle with 3 sets per exercise, moving to 4 sets, and proceeding to 5 sets from week to week. If the athlete tolerates the training well, they can deload as normal and initiate the next training cycle at a higher baseline and progress from 4 sets to 5 and then 6. It is evident that the MRV is not a measure that can be determined in a single training cycle. Determining and reevaluation of the MRV is periodic on the scale of macrocycles rather than microcycles or training days. Changes in MRV largely parallel dose-response relationships relative to training status. It often takes several macrocycles before the coach can define the boundary between training tolerance and excessive volume (MRV). At some point the athlete may come to a point where performing 4 sets of exercise or activity is challenging, 5 sets is really pushing it, and 6 sets extends past the edge of their recoverability. This is the conceptual MRV, where we conclude a reasonable estimate of how much overloading training the athlete can perform without going too far.

We discussed that performance decrements are the quintessential markers for training beyond the MRV; however, performance markers alone can be misleading without supporting evidence. This also applies to the exclusive use of biomarkers or perceptive/psychological indices. For this reason, the authors recommend using three measures: one performance, one biological- or physiological-based marker, and one psychological- or perceptive-based marker for identifying the effects of accumulated fatigue in not only determining the MRV but also preventing nonfunctional overreaching and overtraining syndrome. This represents the essence of allostatic management. The recent textbook *Recovery for Performance and Sport* by Hausswirth and Mujika (2013) outlines many variables that can be used as part of an athlete testing and monitoring program specifically for allostatic management. The authors recommend the following variables that can be used (but are not limited to): heart rate variability, sickness, quantity of sleep, quality of sleep, perception of fatigue, perception of effort or rating of perceived exertion (RPE), feeling of helplessness, testosterone to cortisol ratio (T:C ratio), blood creatine kinase levels, performance decrements, and desire to train.

When considering allostatic management and recovery modalities, we can think of training within the MRV as the prerequisite step before pursuing additional recovery resources. Chronic undertraining is not limited by recoverability, and chronic overtraining cannot be simply ameliorated by pursuing supplemental recovery methods. Allostatic management and recovery operates on a hierarchy, where training within the MRV takes precedence before other strategies. The focus is on utilizing as much overloading training as possible while improving performance to control allostatic load. Once the MRV is established, we proceed into preplanned strategies, which represent the essence of our periodization concept. We use periodization to incorporate directed variation and phase potentiation into an annual plan, along with planned overreaching and active recovery periods through most training phases. Once these are mapped out, we move into concurrent strategies, which include things like nutrition, planned relaxation and stress management, psychological techniques, and therapy to help control fatigue throughout the training process. Finally, we move into posttraining recovery strategies, which are not necessarily meant to be performed continuously throughout the annual plan but can be used selectively to help get athletes back on their feet quickly when they need to perform. This hierarchy is outlined in figure 8.1.

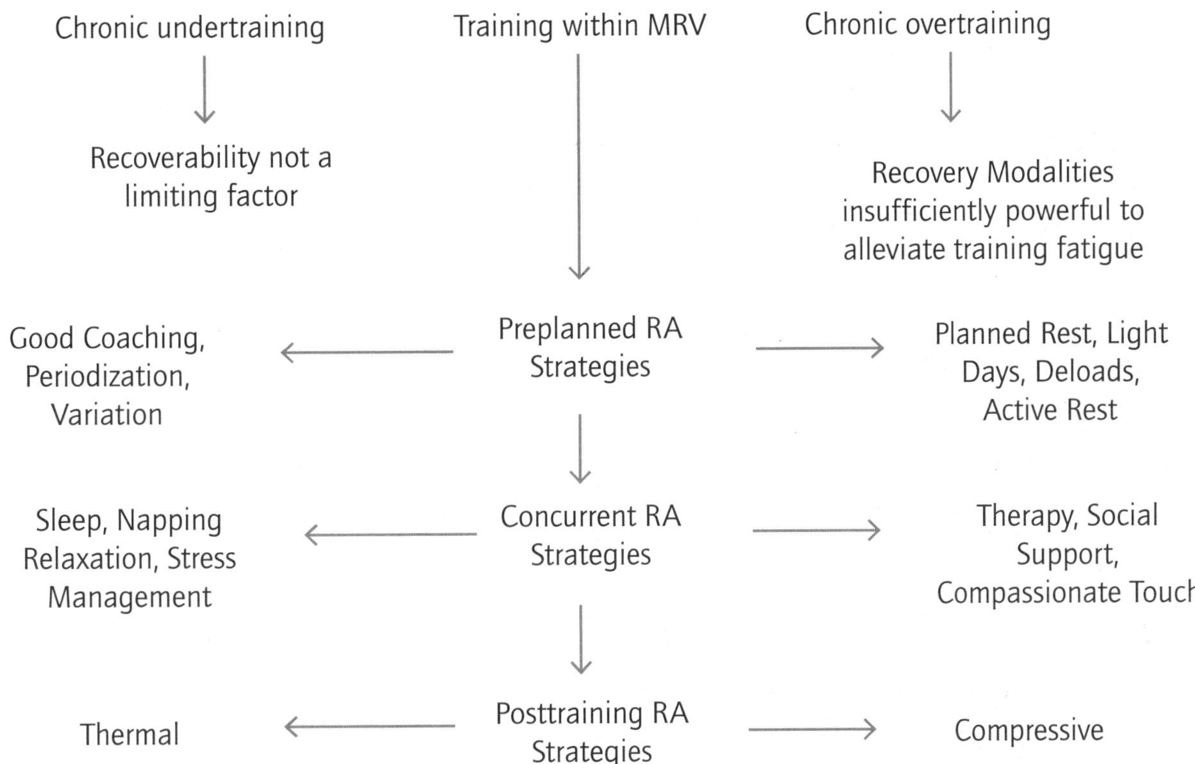

Figure 8.1 Planned approach to allostatic management using MRV (maximal recoverable volume) RA (recovery-adaptive)

NUTRITION CONSIDERATIONS

In general, nutritional variables are not particularly powerful for monitoring and prevention of fatigue. However, nutrition variables act as a supplement to validate data from performance, physiological, and psychological measures. Of particular note are appetite (or lack thereof) and bodyweight. Appetite has been used as an indicator of fatigue, and the results are actually semi-counterintuitive to many laypersons. Typically, when athletes have accumulated chronic fatigue and are moving into nonfunctional overreaching and/or overtraining syndrome, appetite and desire to eat typically drops significantly, often resulting in poor dietary compliance and its associated negative effects. Although hunger can be highly variable day to day, consistent downward trends in appetite can be an indicator of fatigue accumulation.

Similarly, bodyweight fluctuations, particularly weight loss, can also be an indicator of a fatigued state. As previously mentioned a loss of appetite is commonly associated with accumulated fatigue, and accordingly if the athlete is not eating sufficiently, they can suffer the ill effects of hypocaloric dieting. Another factor to consider within bodyweight is dehydration. If the athlete is not replenishing lost body water sufficiently after exercise sessions, they can slowly start to dehydrate over time, which will also have a negative effect on performance and recovery. Unfortunately, bodyweight has a high degree of day-to-day variability even under favorable conditions, so the best assessment of bodyweight measurements is through trend analysis. Although daily measurements can be useful, the authors recommend biweekly assessments of bodyweight measured under consistent identical conditions to properly assess changes in bodyweight. If the trend throughout the week indicates the athlete is losing weight (when they are not scheduled to do so), this may be the result of not eating enough, depressed appetite from fatigue, and/or dehydration. This data by itself may not be representative of the athlete's current state; however in conjunction with data from performance, physiological, and psychological measures, it represents a meaningful component of fatigue assessment.

PSYCHOLOGICAL ASPECTS OF FATIGUE AND OTS

Numerous psychological and perceptive indices have been associated with nonfunctional overreaching and OTS and are an integral part of the monitoring process. Since psychological aspects of fatigue concerning signs and symptoms have been previously mentioned, OTS will be the focus. Mood changes are an early and sensitive indication of OTS. Psychological factors associated with overtraining syndrome are decreased self-esteem and motivation to workout, difficulty concentrating during practice/school/college, general apathy, feeling of sadness, fear of competition, easily distracted during training tasks, "giving up" when the going gets tough, and finally depression and chronic fatigue (Fry, Morton, & Keast, 1991).

To be able to evaluate and diagnose the above symptoms, the coach can use observational methods on the athlete's reaction and behavior during training. Similarly, the athlete can use a self-report to evaluate his/her response to the training load. For example, the athlete is asked to evaluate himself/herself before/after the training session regarding mood, energy, motivation, and emotional state on a 5-point Likert scale. At the end of the training week, the athlete can analyze his/her responses throughout the week. Based on practical experience and interaction with the athlete/coach, there are general and specific signs of OTS. The general symptoms of OTS are fast and irregular fatigue, decreased level of work capacity, sleep disturbance, absence of vitality, and infrequent headaches. In addition, specific symptoms of OTS can be seen as a continuous process. Initially, the athlete may experience episodes of nervousness, which may be characterized in caprice, unstable mood, muscle pain, and nervousness. Later, continuous nervousness is characterized by long-term anger, blaming others, restlessness, pessimism, and instability in training. Finally, depression, anxiety, low self-confidence, high sensitivity, and low motivation and coping abilities can occur.

In cases of irregular behavior and reaction, it is possible to use more accurate psychological diagnostic tools to measure OTS: profile of mood state (POMS: McNair, Lorr, & Droppleman, 1971) and psychomotor performance (i.e., reaction time). The POMS measures six transitory emotional states: tension, depression, anger, vigor, fatigue, and confusion. Overtraining athletes usually show an inversed "iceberg profile" in their POMS's result, which means the negative states of depression, anxiety, fatigue, confusion, and tension become elevated, whereas vigor is decreased. In addition, psychomotor performance may be impaired in athletes with OTS. This may be revealed by reduced or impaired reaction time on motor tasks (Nederhof, Lemmink, Zwerver, & Mulder, 2007).

The diagnosis of OTS can often be only made retrospectively; therefore it is important to focus on prevention OTS by monitoring training load and using recovery means. In addition, early diagnostic of psychological symptoms of OTS is a significant part of preventing overtraining.

TREATMENT MODALITIES

Allostatic management involves several modalities that can be tailored to each sport. Significant emphasis must be placed on planning structured recovery sessions into the integrated periodization model. There are a multitude of methods at the practitioner's disposal to enhance recovery rate after training. In this scope, we use the term *treatment* to characterize the relationship between fatigue and allostatic load in a controlled manner. This section will narrow down the most practical and effective treatment modalities.

Passive recovery assumes high rank among all treatment modalities. Sufficient sleep is a primary moderator of allostasic load and greatly enhances recovery from training. Research has shown that when sleep is altered at a chronic level, almost all performance parameters decrease, that is, aerobic, anaerobic, and mental capacity (Tuomilehto et al., 2017). Although sufficient sleep has considerable individual variation, as seen in US special operators undergoing intensive training, athletes should strive to get at least 6–8 hours of sleep per night consistently throughout the week (Clayton, Brown, Burrell, & Matthews, 2011). Additional sleeping periods during the day are indicated for no more than 15–30 minutes to prevent circadian reset and sleep inertia. Complete cessation of training days should be used sparingly in favor of active recovery. There is no amount of food, supplements, performance-enhancing drugs, recreational drugs, or any other factors that can make up for lack of sleep.

Another powerful form of passive recovery is practicing acute stress management and relaxation. Some scientists and coaches (including some of the authors) speculate that it is impossible to differentiate many of the recovery tools currently used in the field from simple relaxation methods. In the context of allostatic management, the concepts mentioned in this section and their application to athletics boils down to two universal ideas:

1. Shifting the athlete from a heightened physical and psychological state to a lowered physical and psychological state
2. Managing emotional responses and limiting psychological efforts during nonessential times

Stressors, both physical and psychological, have been associated with a number of different adverse health effects and are both major components in chronic fatigue. Although the mechanisms are not always clear, carrying a high degree of stress can have a measurable effect on physiology and sport performance, both good and bad. In fact, we rely on our stress responses during training to heighten our senses, improve energy output, and elicit greater physical outputs than we normally could at a rested state. Unfortunately, that same stress response that allows us to "call the lightning" when we want to is inherently catabolic and, if chronically elevated, tends to disrupt many of the recovery-adaptive processes and desensitize the magnitude of response during actual stressful times like training. Elevations in cortisol, catecholamines, and glucagon are excellent for producing energy and shifting the athlete into a physically and psychologically aroused state; however, these need to be systematically decreased once the need for the heightened state is over.

Incorporating stress management and relaxation is often easier said than done. In an ideal world, athletes would be allowed to practice and train and then be off their feet, carefree, the rest of the day with plenty of nourishing meals. This unfortunately is not realistic. In the real world, athletes need to conserve physical and psychological effort for the most impactful times and then utilize unwinding and relaxation techniques during times where an aroused state is not needed. One of the simplest and easiest ways to incorporate this concept in a realistic manner is planned relaxation throughout the day. The best place to start is generally the immediate postexercise periods. At the cessation of a training or practice session, encourage athletes to rest and get off their feet and find something pleasant and enjoyable to do, even if it's just watching videos or messaging their friends. Make sure steps are taken to rehydrate and replenish carbohydrate stores and bring themselves down to a relaxed state for the next 45–60 minutes. This is an essential time for secondary cellular signaling and the resulting cascade, which downstream leads to enhancement of protein synthesis and gene expression from training. If athletes have jobs or school, they will understandably have to be physically and mentally active throughout the day, but once all major business for the day has concluded, they should make an effort to physically and psychologically unwind and relax. Even something as simple as planning 1 hour throughout the day to have uninterrupted relaxation can be a powerful tool in allostatic management.

Similarly, athletes need to be encouraged not to overrespond emotionally to daily stressors. Although stressful things are bound to happen throughout the day, such as getting in trouble at work, fighting with a significant other, receiving a rude gesture while driving home, and so on, the way one responds to such things is a choice: Becoming upset or flipping out over minutia ends up increasing allostatic load over time and blunting the beneficial effects of the stress response during planned times. Athletes must take time to practice daily stress management strategies to avoid becoming overly emotional or upset for things that are relatively unimportant. Maintaining a generally calm and relaxed state takes practice. This is often easier said than done, but athletes can be conditioned through allocated relaxation times and repetition.

Active recovery is another effective modality that uses light exercise loads to facilitate and enhance recovery. Active recovery modalities such as taking light days, unloading (deload) periods, active rest phases, peaking and tapering phases, and other similar methods are among the most well-documented methods of alleviating fatigue, making them an incredibly powerful tool in the allostatic management tool kit. Interestingly enough, active recovery methods such as taking a light day can often have a greater effect on alleviating fatigue than complete rest. The following illustrates some of the most practical ways of including active recovery methods into an annual training plan.

Planned and impromptu light training days are one of the simplest and most effective forms of active recovery. In short, a light training day is a training day in which there is a distinct reduction in the training volume and possibly a small or modest reduction in the relative training intensity with the goal of alleviating acute fatigue. Light days seem to have a positive effect on enhancing blood flow to the working muscle, thus enhancing nutrient uptake and waste removal, and possibly help facilitate the signaling cascade for the recovery and adaptive processes without accumulating any significant amount of fatigue. Typically, athletes who take a light training day usually go into the session feeling fatigued and worn, but they finish the session feeling slightly rejuvenated and less beat up. There are many options and opportunities for a light training session; however the key is a mandatory reduction in the training volume and relative intensity. Simply switching up the activities and continuing voluminous training does not achieve this effect.

The amount of volume reduction needed is often determined by the athlete's training level and magnitude of accumulated fatigue, but for most a reduction of approximately 30–50% of training volume is appropriate for a light day. This applies to not only strength and conditioning activities but also sporting activities. Instead of performing high bar squats for 3×10, an athlete might perform the same exercise at 2×10. If a normal overloading Rugby practice lasted for 90 minutes, a light session would only last 45 minutes. The relative training intensity can also be reduced during a light session for most activities; however, it is worth noting that for some highly technical activities, maintaining the proper "feel" and familiarity with the technique may require a maintenance of the training intensity. Generally, it is recommended that the relative intensity be lowered about 15% of the normal overloading session for the week. For example, if an athlete was high bar squatting 3×10 at 85% of their 3×10 best, they might squat at 2×10 at 70% of their 3×10 best for a light day. Similarly, for the normal 90-minute Rugby practice, the 45-minute practice can consist of the normal warm-up, some basic skill work like passing and catching, and some light drills, but mainly limit the amount of contacts and maximal efforts.

It is important to note that much of the research on light active recovery is done using cardiovascular exercise such as walking, cycling, and swimming. For normal exercisers this strategy is fine; however, for athletes a strong case can be made for always continuing to perform normal training activities not only to target the prime movers but also to reinforce the skills and technique of the sport. Just because the training session is light, this does not mean the athlete cannot benefit from continuing to practice and enhance training and sport skills. Practically, this means that the squat technique should still be crisp, passes and catches not dropped or fumbled, and running and sprinting mechanics intact. Many modalities can elicit the effect of light active recovery; however, the authors recommend maintaining specificity and continuing to develop the training and sport skills needed to be successful rather than adopting new modalities for the purposes of active recovery.

Similar to light days, unloading phases (commonly referred to as deloading phases) seek to alleviate training fatigue but on a more comprehensive scale. Simply put, the deload phase is a series of light training days linked together into one microcycle, typically programmed after planned overreaching periods. The deload seeks to reduce both acute and accumulated fatigue from weeks of continued intensive training. This applies to not only physical stressors such as substrate depletion, microtrauma, and unfavorable nervous and endocrine alterations but also psychological factors and cumulative allostatic load. Like the light day, the deload phase is marked by a substantial reduction in the overall training volume. This reduction is dependent on the athlete's training level and the magnitude of accumulated fatigue. The range of decrease is approximately 30–60% of total overloading training volume (across all training sources). A planned reduction in intensity is suggested during these phases to help reduce the psychological strain of preparing for large efforts but also to allow for healing of accumulated microtrauma, which tends to be related to the absolute load or effort. Planned deloading phases should occur once every 4–6 weeks after planned overreaching periods or difficult periods of training. For most athletes this typically lasts for one microcycle.

The active rest phase may not be of particular importance to the general populace, but it is an integral component of an athlete's annual plan. Long periods of competitive training phases take their toll on the body and the mind, resulting in substantial allostatic load. Subsequently, it is commonplace for many athletes to experience burnout or become disinterested in the training process after months of training. The active rest phase is a period of complete healing and recovery. During this phase, the athlete is encouraged to remain active and maintain fitness through less structured and rigorous training and take a break from the rigors of their sport. The athlete is encouraged to seek out activities that will maintain basic fitness characteristics without being overly specific to their sport. This phase is typically programmed at the end of the competitive phase before transitioning into a new period of training. The duration of this phase is highly dependent on the amount of psychological stress carried by the athlete and current athletic abilities. For young or relatively new athletes, this effect can be elicited in as little as one week, similar to a deload week. Intermediate athletes appear to benefit from about 2–3 weeks of active rest, and highly trained athletes, especially large strength athletes, require 3 weeks or more to achieve this effect. The majority of athletes' active rest phases are recommended for about 2 weeks at a time, 1–3 times per year, with an adjustment buffer for the needs of the individual. Upon completion of the active rest phase, there should be a marked increase in desire to train and excitement to begin the next training cycle.

Although there are supplemental recovery modalities available, which can have a positive effect on enhancing recovery, practitioners should be cautious before relying too heavily on such methods as they are not sufficiently powerful to alleviate the effects of nonfunctional overreaching and overtraining syndrome. Often supplemental modalities come at the cost of blunting the adaptive processes from training. Of particular note are thermal and compressive modalities. Both modalities have been shown to be effective at promoting recovery from training and competition, but they also result in a lower magnitude of response from training due to effects associated with reduction of the natural inflammatory response. These methods are best utilized during highly competitive periods where performance is favored over long-term adaptation. These should be minimized during periods away from training, such as a general preparatory phase, and may not be required at all for developing and/or younger athletes who should focus more on establishing good training habits and determining their tolerance to normal training.

NUTRITION INTEGRATION

One of the most powerful and practical alleviators of fatigue from a nutritional standpoint is simply calories, period. A hypercaloric state has a profound effect on alleviating fatigue from hard training. Much of this fatigue can be explained by substrate depletion and energy balance. One of first treatments for coaches who suspect their athletes are overly fatigued or tipping the boundaries of nonfunctional overreaching and/or overtraining syndrome is to balance energy levels by either reducing training load, increasing calories, or including some combination of both factors. This is especially true for athletes who are experiencing nonplanned weight loss. An increase of approximately 250–500 calories per day is the first place to start; however, this may require a more substantial increase of approximately 500–1,000 calories per day if the athlete is undergoing extremely voluminous training loads.

In the above example, we are considering the guidelines from chapter 2 on macronutrients. We do not necessarily want to add 500 calories of peanut butter or steak, though this still would present a positive effect. If allostatic management and recovery are the primary issues, then a large portion of those extra calories, as well as total calories, should be coming from carbohydrate. Protein is obviously important for muscle maintenance and recovery, but in the scope of recovery, it will take a second seat behind carbohydrate. Timing of carbohydrate intake will also play a more important role in allostatic management, as the majority of daily carbohydrate consumption should be near the pre, intra, and post exercise periods. More glycemic carbs may also be introduced at this time, as they are generally easier to consume, more palatable, and can slightly enhance the rate of glycogen synthesis.

A typical field application of this knowledge proceeds in the following fashion. The athlete begins to present small but consistent decreases in performance from week to week. They report feeling tired, sluggish, and possibly anxious with a downward trend in bodyweight starting to form. The coach's first intervention can be to instruct the athlete to take a few light training days and increase baseline calories by about 250–500 calories per day. This comes primarily from carbohydrate sources, equating to roughly 63–125 g of additional carbohydrate per day, or about 1 or 2 extra bagels throughout the day. If the athlete begins stabilizing bodyweight and returning to baseline performance, then no additional treatments may be necessary, though continual monitoring is required. If symptoms do not improve, the coach can instruct the athlete to move into a deloading phase or significantly reduce their training loads and simultaneously continue to increase daily calories. If symptoms do not improve at this point, a more thorough investigation into the athlete's situation is warranted.

PSYCHOLOGICAL INTEGRATION

To avoid OTS and optimize performance, training recovery should be systematically planned and implemented. Therefore, if an athlete is overtrained as a result of physical load, strategies such as nutrition, massage, passive rest (i.e., no training), and active rest (i.e., low intensity training in a different sport, stretching) should be used. On the other hand, if the overload results from psychological factors, strategies such as relaxation (e.g., progressive muscle relaxation), relaxing breathing, visualization, dissociation, and thought management strategies ought to be implemented. In team sport improved team cohesion, interpersonal relationship, and emotional management should be used as recovery activities (Elbe & Kellmann, 2007; Weinberg & Gould, 2015).

More specifically, the psychological strategies should be used systematically and be part of the training regime/plan. For example, a few minutes of muscle relaxation between exercises, and approximately 10–15 minutes of muscle relaxation at the end of training, together with relaxing breathing, have to be integrated in practice. At the end of the training week, the athlete should perform relaxation together with focusing on positive indicators (e.g., coping with training load, positive thoughts) of the past training week. In addition, mindful walking in the nature can serve as refreshing/recovery means. In team sport, players are sometimes engaged in 15–30 minute light jogging after training and listen to favorite music on the way back as a recovery mode. Finally, and above all, the most significant factor for prevention of OTS, in one hand, and recovery from OTS, on the other hand, is the athlete's self-regulation skills (Hodge & Kentta, 2016).

SLEEP DISORDERS

Sleep disorders are often overlooked but play an integral role in health and athletic performance. Restorative sleep is essential for optimal recovery and allostatic load reduction. Athletes incur sleeping problems at various points in their careers, especially near competition and during travel. Most coaches can easily identify athletes with major sleeping problems through direct observation. Consistent vigorous physical activity of athletes has been shown to facilitate circadian rhythmicity; however, some athletes develop sleeping disorders for a multitude of reasons. Current research in athletic sleeping disorders is limited, but many important concepts can be gleaned from the existing studies and practical experience. Athletes must have more sleep than sedentary counterparts or recreationally active enthusiasts.

Sleep deprivation associated with sleep disorders carries several negative consequences. Impaired quality and quantity of sleep has been shown to cause mental lapses, slower cognitive ability, impaired motor learning, impaired memory, decreased vigilance, loss of focus, and a decline in reactive ability (Davenne, 2009). It is important to consider sleep and activity as physiological factors that normally alternate and mutually preclude each other.

The prevalence of sleep disorders is usually underestimated. One study of 107 professional ice hockey players evaluated quality of sleep and prevalence of sleep disorders (Tuomilehto, et al., 2016). The researchers performed an exploratory observational 1-year follow-up study using a questionnaire sleep assessment along with counseling, as needed, and polysomnography with individual treatment plans. Interestingly, one out of four players revealed major sleeping problems. This provides a reasonable estimate that warrants further attention of practitioners involved with team sports.

TREATMENT OF SLEEP DISORDERS

Standard medical treatment can be problematic because the cause of a sleep disorder is usually the product of multiple factors. For the coach, this is often too complex to address. If the sleeping issue is severe, referral to a medical doctor with experience with sleep disorders is preferred. However, moderate sleeping disorders can be identified through assessments such as the Athlete Sleep Screening Questionnaire (Samuels, James, Lawon, & Meeuwissee, 2016). This screen was designed to assess and manage sleep in elite athletes.

The magnitude of sleep disorders is beyond the scope of this book. However, once a problem is confirmed using a modern questionnaire, counseling and further investigation is warranted. The first tier of treatment would be counseling on proper sleeping habits and how to adjust schedules to meet demands of daily life. This is especially true for student athletes who often get far too little sleep. Counseling and questionnaires will likely uncover covert behaviors that can be the "real" cause of the problem.

If the sports professional cannot ascertain the problem, a medical doctor can perform a short sleep study with polysomnography. This process can uncover a wealth of information and provide the data needed to create an individualized treatment plan for the athlete. Keep in mind, the whole process of integrated periodization involves effective communication across many disciplines. It is important for all professionals involved to be aware of the problem and communicate vital information to the team or consulted physician.

PSYCHOLOGICAL ASPECTS OF SLEEP DISORDERS

Based on sport psychology consulting knowledge, when an athlete experiences sleeping difficulties (i.e., excluding clinical/medical condition), there are general recommendations for coping with the situation. Initially, the last training session and dinner time should end up at least 2 hours before initiation of sleep. Short light walking (e.g., 20–30 min), listening to favorite relaxing music, reading books, performing progressive muscle relaxation with relaxing breathing, and imagery of nature walk or favorite relaxing scenes are important methods for good night sleep. However, specific recommendations can be prescribed to the athlete based on individual case characteristics, personality qualities, and dynamics of the situation.

SUMMARY

This chapter presented a holistic approach termed *allostatic management* to integrate exercise physiology, nutrition, and psychology in accordance with the principles of periodization. The intent was not to replace the fundamentals of the training process but to broaden the practitioner's arsenal in the quest for optimal performance. Both homeostasis and allostasis play a key role in our understanding of how short-term stimuli transfer into concrete, long-term, beneficial adaptation. This is achieved through both "stability through constancy" and "stability through change" with the ultimate goal of increased sports performance.

Pertinent relationships between fatigue, overreaching, and overtraining were explored with emphasis on physiology, nutrition, and psychology. The allostatic paradigm offers a system to prevent fatigue, overtraining, and increase sport performance using specific methods that can be feasibly incorporated by coaches and strength training specialists. As sports professionals, we must seek a higher understanding of how physiological, nutritional, and psychological principles can be used more effectively in our quest for athletic excellence. This process is realized through integrated periodization.

Chapter 9

TAPERING AND PEAKING FOR COMPETITIONS

Tudor Bompa, PhD / James Hoffmann, PhD / Boris Blumenstein, PhD / Iris Orbach, PhD

Athletes train for many months for the sole purpose of testing their athletic potential against their opponents in an organized athletic environment. During training, athletes are exposed to many types of stressors, from the challenges of perfecting their skills to physical, social, and acute levels of neuropsychological fatigue. While during the months of training athletes have to overcome multiple challenges during competitions, they encounter the Sisyphean tasks of achieving peak performance. To successfully reach high performance at the desired time, coaches have to design specific tapering strategies that will facilitate peaking.

TAPER AND PEAKING DEFINED

Taper is an essential short phase of a well-planned training program designed to induce peak performance for the main competitions of the year. A well-planned taper has the major scope of reducing all the stressors athletes are exposed to in order to remove, prior to a major competition, the accumulated levels of fatigue acquired during the preparatory phase of the annual plan. Since during taper the intensity, volume, and frequency of training are progressively reduced, athletes can also reach a relaxing mode, use best nutrition plans to restore depleted energy reserves of the body, and, as a result, reach *supercompensation.* Therefore, supercompensation has to be seen as a *sine qua non* condition, to reach a state of physical and psychological arousal prior to an important competition. Since fatigue is a fast-responding, direct effect of training, the scope of tapering is to dissipate all the physiological and psychological elements that may interfere with athletes' goal of reaching peak performance during competitions.

Peaking, on the other hand, refers to the time when an athlete reaches the state of *readiness* for competitions, when peak performance is achievable. Therefore, tapering triggers a chain-effect:

taper > supercompensation > peak performance

Figure 9.1 illustrates the progression from training during preparatory phase, the competitive phase, leading to the precompetition taper and peak performance at the time of competition. This example represents the activities planned for an individual sport with one competitive phase only. In the case of sports with several competitive phases and peaks, the proposed activities/methods have to be condensed into shorter preparatory and competitive phases.

Training phase	Preparatory phase	Competitive phase		
		Training for competitions	Precompetition tapering	Main competition peaking
Duration	Months	Weeks/months	7–10 days	1–3 days
Activities	Technical and tactical Improve/increase adaptation via specific training, based on the dominant energy systems and dominant physical abilities	Technical and tactical training for competitions Maximize training and adaptation as per the specifics of the sport and dominant abilities Maximize athletes' capacities to tolerate fatigue and perform under highest level of physiological and psychological stress	Readiness for competition Apply physical, nutrition plans and psychological techniques to trigger supercompensation	Peaking Supercompensation trigger peaking
Methods	Increase capacity to tolerate and cope with fatigue Nutrition plans Psychological strategies	Specific training methods aimed at maximizing physiological and psychological capacities to reach highest level of adaptation and performance Nutrition plans as required by the dominant abilities and dominant energy system/s Psychological strategies	Reduce physiological, social, and psychological stressors to facilitate peaking	Apply physiological, nutrition and, psychological techniques to facilitate peak performance

Figure 9.1 The chain-effect of training phases: tapering and peaking for the main competition

Although tapering and peaking are part of the modern vocabulary constantly used by coaches and athletes, you might be surprised to know that tapering strategies were also used by the Greek Olympians some 2,500 years ago. One has to be quite humbled to know we weren't the first ones! Others preceded us on the stage of academia, science, and methodology! One of them, Phylostratus, the Greek physician and scholar, exemplified the *tetra system* (a four days training plan) used by athletes as a tapering strategy prior to the ancient Olympic Games:

Day 1: short and energetic

Day 2: exercise intensely

Day 3: relax and revive the activity

Day 4: perform moderate exercises

How intriguing to find out, after all, we aren't the first to use tapering as a method to trigger peak performance! However, and in spite of that, modern knowledge of training has never reached the state of development than the beginning of millennium 3! That the information we pride ourselves with have evolved from scientific findings, research and development, but also from coaching expertise. No doubt, the future belongs to inquisitive minds, to those who dare to challenge the status quo! And this is quite rewarding to know!

TAPERING MODELS

The scope of any tapering models is to reach a state of *readiness* for competition and, as a result, to optimize athletic performance. This is achieved as a result of reducing the intensity, volume, and frequency of training in a progressive manner. Mujika (2009) reviews the three most popular tapering models used in different studies or by most coaches:

1. *Linear* model implies a continuous, progressive, but linear reduction of training demand.
2. *Step* model is in fact a reverse model used for increasing the load in training. Step taper, therefore, is used to decrease all training stressors in smaller, progressive steps.
3. *Exponential*, or the rate of decreasing volume, intensity, and frequency of training to induce peak performance, has two variants: *slow* and *fast decay*.

From analyzing available research studies regarding taper it appears that the most efficient tapering model has been found to be the *exponential, fast decay,* resulting in a 4%-7.9% improvement, whereas the *step* model has produced a nonsignificant improvements: 1.2–1.5% (Banister, Carter, & Zarkardas, 1999; Zarkardas, Carter, & Banister, 1995).

HOW TO MANIPULATE TRAINING TO INDUCE TAPER AND PEAKING

In a colloquial sense, training is a manipulation of methods intended to achieve highest physiological and psychological adaptation to achieve desired readiness for competitions. By the same token, when the highest adaptation occurs, ideally before main competitions, coaches manipulate tapering models to trigger supercompensation and, as a result, facilitate peak performance. Reduction of training demand, volume, intensity, and frequency normally facilitate the removal from the body and mind of previously accumulated fatigue and facilitate supercompensation.

Invariably, coaches' goal during tapering is to reduce the most fatiguing training activities in the selected sport. A brief analysis of the three main energy systems follows:

1. The *phosphagen* energy system is dominant in short-duration, high-intensity, and maximum-concentration types of activities. The majority of training activities have a duration of 5–15 seconds long. Energy is supplied by the phosphagen system (ATP-CP). Sports that belong to this category are speed-power events, such as field events in track and field, sprinting, gymnastics, baseball, elements of martial arts, fencing, weight lifting, and so forth. For these sports the *fatiguing element* is high-volume, longer-duration, and aerobic type of training activities.

2. *Glycolitic* system supplies the necessary energy for sports of 20-second to 2- or 3-minute duration such as prolonged maximum speed events (400–1,500 m in track and field, 100–200 m swimming, speed skating), combat, martial arts, racquet sports, and so on. Considering the specifics of these sports, in the early part of the activity, phosphagen is taxed, followed by glycogen, and at the end of the race/activity, the oxidative system also contributes to the total energy used in these sports. In spite of the fact that all three energy systems are taxed in specific percentages, the most fatiguing element for the sports in this category is oxidative, long-duration, nonstop, and steady physical activity.

3. *The oxidative system* is dominant for long-duration, lower intensity sports lasting from 3–5 minutes to hours of nonstop supply the energy. In the initial stages of activity, energy is supplied by glycogen, but as the nonstop duration increases, the body relies on the energy produced by fatty acids and, ultimately, by protein sources. Most sports, from team sports to marathon, triathlon, Nordic skiing, road cycling, and so on, rely in a smaller or larger percentage on the energy produced by the oxidative system. Most athletes involved in long-duration, nonstop activities have a difficult time to cope with or tolerate the fatigue induced by high-intensity type of training.

Special note:* During taper *avoid the most fatiguing, nonspecific activities in your sport! Equally true, never expose your athletes to what they are not used to!

REDUCTION OF TRAINING LOADS TRIGGERS SUPERCOMPENSATION

Since supercompensation facilitates peaking, there are three strategies that one can employ to reduce training loads during taper. Training programs during taper, however, should aim at *removing fatigue without decreasing fitness level and its negative effect of detraining.* Therefore, the motto of the strategy is this: *reduce what is fatiguing*!

When designing a tapering plan, you may consider the following strategies of reducing intensity, volume, and frequency of training:

Reduce Intensity

- During taper, train activities that maintain adaptation (Bosquet, Monpetit, Arvisals, & Mujika, 2007).
- Although high intensity should be progressively reduced, you should still maintain the types of training that give the athlete the *feel of maximum speed* (McNeely & Sadler, 2007). It also gives the athlete the confidence of training what they are used to (Bosquet et al., 2007).
- The feeling of *being energized*, rather than fatigued, is very essential for the psychological well-being of the athlete (Bosquet et al., 2007).
- When training high intensity, reduce its duration, but increase rest interval to avoid fatigue.

Reduce Volume

- To optimize adaptation, decrease volume of training by 41–60% of pretaper training volume (Bosquet et al., 2007).
- Maximal performance gains were obtained with a reduction of training of 41–60% of the pretaper training volume (Bosquet et al., 2007). However, gains in performance are also possible with smaller or bigger training volume (Mujika, 2009).
- Adaptation and performance gains are sensitive to reduction of training volume (Rietjens, Keizer, Kuipers, & Saris, 2001).
- During pretaper and taper, eliminate all physiological, psychological, and social stressors.

Reduce Frequency

- Reduced training frequency by 50% resulted in performance improvement (Dressendorfer et al., 2002).
- Researchers have found that maintaining the daily frequency of training versus resting every third day resulted in better performance: 1.93% versus 0.39 in a 800 m race (Kubukeli, Noakes, & Dennis, 2002; Mujika et al., 2002). The authors attributed lower results to the second group to a "*potential loss of feel*" specifically because the training volume has been decreased.

- It seems that decreasing training frequency is not as significant in performance as much as decreasing the volume and intensity per training session (Bosquet et al., 2007).
- Most coaches have great imagination to maintain frequency athletes are used to but decrease both volume and intensity during taper.

BENEFITS OF TAPER

The methodology of tapering has been demonstrated scientifically to be beneficial for all sporting activities: from oxidative energy system to team sports. A condensed version of scientific data is presented below.

Taper for the Oxidative Energy System

- As a result of applying taper methodology with a reduction of 50%, muscle oxidation has been increased (Neary, Martin, & Quinney, 2003)
- Resting heart rate has been decreased (Dressendorfer et al., 2002)
- As a result of 7 days taper, the volume of red cells have been increased by 15% (Shepley et al., 1992), a clear benefit for the athletes since this increases the oxygen-carrying capacity during the race
- Increases hemoglobin concentration (Mujika et al., 2000)
- Muscle glycogen concentration in the blood increases by 17–29% after a taper of 4–8 days (Neary et al., 2oo3), a demonstration that more energy reserves are available after taper
- As a result of 7 days taper, VO_2 max has increased by 6.0% (Neary et al., 2003) to 9.1% (Banister et al., 1999; Margaritis, Palazetti, Rousseau, Richard, & Favier, 2003)

Taper for the Neuromuscular System

- Taper-induced improvements of strength and power has been reported by several studies (Raglin et al., 1996; Trappe, Costill, & Thomas, 2001; Trinity, Pahnke, Reese, & Coyle, 2006)
- A 7 days taper increased performance by 7–20% (Izquierdo et al., 2007; Trappe et al., 2001)
- Arm power improvement by 10–12% has resulted in a performance improvement of 4.4%

Taper for the Endocrine System

- Although changes are not always present in the literature, tapering-induced improvements in testosterone, cortisol, or the ratio of testosterone to cortisol (T:C ratio) have been documented (Mujika, Padilla, Pyne, & Busso, 2004).

- It has been assumed that catecholamine hormones would also see an improvement from tapering periods due to their use in identifying and monitoring stress, overreaching, and overtraining syndrome. The literature however has remained inconsistent at best, and although a strong theoretical case can be made for taper-induced positive changes in catecholamine concentrations, more research needs to be done before any conclusions can be drawn.
- Human growth hormone (HGH) and insulin like growth factor-I (IGF-I) have shown improvements from tapering; however at the current time, there is not enough literature support to be conclusive (Mujika, 2009).

Team Sports: A Real Taper Challenge

Discussions regarding tapering for team sports are an incredible challenge. The difference between team and individual sports is immense, mostly because of the number of games a team sport player has to peak for. Equally important is that there is very little information, both scientific and anecdotal. If in individual sports a scientist can isolate specific abilities such as speed, power, and endurance, in team it is difficult to quantify them. The only quantification we have observed recently was the distance soccer players have run during the games of the European Soccer Championships of 2016. Some players were running 11–14 km per game! There are several difficulties we have to consider when analyzing taper methods in team sports are:

- Difficulty to isolate intensity versus volume
- Try to quantify agility, so important in most team sports! How can you quantify in isolation the contribution of strength and speed to the development of agility?
- Playing season for professional players is around 10 months. For many players the duration of the competitive phase in 2016 was 11 months!
- Most players participate on 60–70 games!
- When do they rest/recover prior to the next season?
- When do they have time to train for the next season?
- Exposed to high risk of injuries

Tapering for team sports tends to be more challenging than strength sports like powerlifting and weight lifting simply due to having a greater number of training components that must be accounted for. As previously mentioned one of the essential components to tapering is a major reduction in the training volume, which comes from not just one but all sources. Team sports typically have, but are not limited to, strength training, sport practice, conditioning, and any type of agility, power, and speed components at any given time. All of these sources, not just the weight training, need to be reduced. This will mean progressively less time on the practice field, less time in the weight room, and less time jumping, sprinting, and practicing change of direction (Mujika, 2007).

On a similar note because the duration and stress level of each session is reduced, coaches may be able to consolidate training sessions that were spread out throughout the week into split sessions within the same day to allow for more distinct recovery periods throughout the taper. So instead of splitting weight training sessions from practice sessions throughout the week, which normally may have needed to be done for allostatic management purposes, those may be consolidated into AM/PM sessions to allow for distinct recovery days throughout the week. This may allow for a reduction in the training frequency of some components, allowing for more distinct physical and psychological recovery periods throughout the weeks of the taper.

DURATION OF TAPER

Determining the duration of tapering is not an easy task. You are in an environment where scientific and anecdotal information collide, where traditions and mentalities influences the final decision. Often, duration depends on the state in which an athlete is before starting tapering. If high levels of fatigue, overreaching, has been experienced by athletes, duration of taper has to be longer than normal.

The duration often has to consider the time necessary to dissipate training-induced fatigue. The sport in which an athletes has specialized is often a factor. Sprinters in athletics use a two weeks taper, while rowers traditionally use only 5–7 days. Swimmers often adhere to strong traditions of 4 weeks taper. Bosquet et al. (2007) suggest that duration of 8–14 days seem to be the borderline between positive and negative (detraining) benefits. Boxing, on the other hand, has a strange tradition: bouts of maximum speed punches in 15–30 seconds! This is not a taper strategy but rather an illusion: if you can deliver fast punches now, the same will happen tomorrow during the match!

Most athletes, however, use a duration of 2 weeks to 4–5 days. A prolonged duration might take the athletes to a state of detraining, decreasing the benefits of previously reached positive adaptation (a state of *readiness* for competition) and even *losing the feel* of daily training. Besides everything else, the duration of taper is often an individualized training notion.

Scientific information can certainly add to the data of information needed for coaches to decide own strategy and duration of taper. Mujika et al. (1996) suggest that there is a significant correlation between percentage changes in the testosterone-cortisol ration, and the performance improvement and elimination of accumulated fatigue. This indicates enhanced recovery and the start of triggering supercompensation. Shepley et al. (1992) reported increase in red cell volume-hemoglobin levels as a result of tapering. Thomas and Busso (2005) propose that there must be an optimal compromise during taper between the dissipation of fatigue and the preservation of previously reached training adaptation. Coaches must be careful when they use a longer taper than 10–24 days since the undesirable effects of detraining might affect performance expectations.

DESIGNING A TAPER PROGRAM FOR COMPETITIONS

Scientific and anecdotal information lead to a similar conclusion that athletes' reactions to a standard taper seem to be individualized. Therefore any well-designed taper has to also be addressed to individual athletes. Furthermore, Vollaard and Shearman (2006) suggested that *expectancy* regarding tapering effect may influence the benefit of taper, and as such, a taper plan might not work for all athletes. Logically, if adaptation is individualized so should be a taper strategy. Therefore, it makes practical sense for coaches to consider individual reactions when designing a taper strategy for major competitions.

Taper is a designed plan to induce peak performance, but one might not expect miracles. This is why coaches should not consider tapering anything else but a manipulation of training to facilitate supercompensation. This is why an *integrated program for taper, training, nutrition, and psychology strategies is the best road to reach peak performance.* Below are presented several versions of a tapering strategies (Mujika & Padilla, 2003).

TWO COMPETITIVE BOUTS

In many sports there are several levels of competition before one qualifies for the final. The time between bouts is often quite short, such as in combative, racquet, sprinting events, field events in track and field, and even during tournaments in team sports. This is why coaches have to design a recovery/regeneration programs that may energize the athletes for the next bout. Such a program must consider nutritional and psychological strategies. The following describes two recovery methods between two competitive bouts (Bompa & Haff, 2009):

1. A scenario for two competitive bouts separated by 20 minutes of recovery. During the recovery period the athlete is involved in 3.5 minutes of active rest (50%), 7.5 minutes of massage, and finally 3.5 minutes of active recovery (50%).
2. A post-training recovery program lasts 30 minutes in which the athlete is involved in 15 minutes active recovery and 15 minutes of water immersion together with consumption of carbohydrate protein beverage.

Nutrition for multiple bouts within the same day or a short amount of time generally revolves enhancing glycogen repletion through nutrient timing and food types. Although protein is still important, structural damage to tissues is less likely to be a limiting factor to performance than simple substrate depletion. Additionally there is not a sufficient amount of time to regenerate significantly damaged tissue; thus carbohydrate must be emphasized in order to maintain subsequent performance.

Athletes should generally stick to high glycemic carbs such as sports drinks during and immediately following competition. It is paramount that they consume these high glycemic carbohydrates at a rate of at least 1 g per kg of bodyweight per hour following the first competitive bout. After about 1 hour athletes can switch to more moderate glycemic carbohydrates and lean protein sources such as greek yogurt, chocolate milk, kids cereal, fat-free gummies, and fig newtons. It is also important to find foods that sit well with the athlete and will not cause undue gastrointestinal distress. For example, drinking too much sport drink may cause the athlete to feel bloated and full of water, whereas frosted kids cereal may go down more easily and comfortably. Similarly high fat and fiber foods may cause them to be too full and feel sluggish or nauseous before competition. This may require some trial and error on from the athlete, which should generally be rehearsed prior to any major competitions.

Motivation is an essential element in multigames tournaments. Players are challenged and fatigued, and, yet, in spite of that, they have to be up for the next game. In the case of professional sports, the mighty dollar is a determinant motivation, but for junior and college athletes, athleticism and excelling in the game is the typical motivation. The following examples illustrates different types of competition plans for team sports/other sports with similar programs.

Figure 9.2 is a standard competition microcycle with two competitions per week, more typical for individual sports. Please note the suggested activities for each day of the week, including the precompetition two-day unloading (reducing the daily load of training) program intended to trigger supercompensation for the Saturday competition. In team sports there may be two games per week, spread out over the week or both during the weekend. A two-days per week games is illustrated by figure 9.3, while figure 9.4 suggests the types of activities to plan between the games. Please note that the program for the first day following each game is dedicated to rest/recovery/ regeneration. In other words, rest and recover before you'll start to train again. Please remember that *rest has to be planned as any other type of training*!

Days of the Microcycle	Training Demands					
	Recovery 0	Very low ‹ 50%*	Low 50-70%	Medium 70-80%	High 80-90%	Very high 90-100%
Saturday	Competition					
Sunday	Rest					
Monday	Regeneration					
Tuesday	Technical Training					
Wednesday	Technical Training					
Thursday	Unloading					
Friday	Unloading					
Saturday	Competition					
Sunday	Rest					

*% - percentage of maximum performance

Figure 9.2 Weekly competitions microcycle with two competitions per week (individual sports)

<table>
<tr><th rowspan="2">Days of the Microcycle</th><th colspan="6">Training Demands</th></tr>
<tr><th>Recovery
0</th><th>Very low
‹ 50%*</th><th>Low
50-70%</th><th>Medium
70-80%</th><th>High
80-90%</th><th>Very high
90-100%</th></tr>
<tr><td>Monday</td><td colspan="3">Regeneration</td><td></td><td></td><td></td></tr>
<tr><td>Tuesday</td><td colspan="4">Tactical Training</td><td></td><td></td></tr>
<tr><td>Wednesday</td><td colspan="6">Competition / Game</td></tr>
<tr><td>Thursday</td><td colspan="4">Regeneration</td><td></td><td></td></tr>
<tr><td>Friday</td><td colspan="5">Technical / Tactical Training</td><td></td></tr>
<tr><td>Saturday</td><td colspan="3">Unloading</td><td></td><td></td><td></td></tr>
<tr><td>Sunday</td><td colspan="6">Competition / Game</td></tr>
</table>

*% - percentage of maximum performance

Figure 9.3 Competitive microcycle with two games in one week (team sports)

<table>
<tr><th rowspan="2">Days of the Microcycle</th><th colspan="6">Training Demands</th></tr>
<tr><th>Recovery
0</th><th>Very low
‹ 50%*</th><th>Low
50-70%</th><th>Medium
70-80%</th><th>High
80-90%</th><th>Very high
90-100%</th></tr>
<tr><td>Monday</td><td colspan="3">Regeneration</td><td></td><td></td><td></td></tr>
<tr><td>Tuesday</td><td colspan="5">Technical / Tactical Training</td><td></td></tr>
<tr><td>Wednesday</td><td colspan="5">Technical / Tactical Training</td><td></td></tr>
<tr><td>Thursday</td><td colspan="4">Unloading</td><td></td><td></td></tr>
<tr><td>Friday</td><td colspan="3">Unloading</td><td></td><td></td><td></td></tr>
<tr><td>Saturday</td><td colspan="6">Competition / Game</td></tr>
<tr><td>Sunday</td><td colspan="6">Competition / Game</td></tr>
</table>

*% - percentage of maximum performance

Figure 9.4 Competitive microcycle with two games in one weekend (team sports)

In many instances there are two days competitions during the weekend. Figure 9.4 illustrates post- and pregame activities, where rest/regeneration and unloading to taper for the weekend games are standard activities. Finally, figure 9.5 illustrates the activities planned for a week-long tournament. The only type of training planned between two games/competitions is low intensity tactical training to ready the athletes for the team they'll meet during the next game.

Days of the Microcycle	Time	
	A.M.	P.M.
Monday	Game	
Tuesday	Regeneration	Tactical Training
Wednesday	Game	
Thursday	Regeneration	Tactical Training
Friday	Game	
Saturday	Regeneration	Tactical Training
Sunday	Game	

Figure 9.5 Week-long team tournament microcycle

Often, between the national league games and a major international competition, such as continental or World Championships/Olympic Games, there are few weeks of training camp/transition from national to international competition. Figure 9.6 proposes a four-microcycle training between national league and an international tournament.

Microcycles	1	2	3	4	5
Scope of training	Rest, recovery/regeneration Remove fatigue Nutrition plan Design psychological strategies necessary to apply in each microcycle as per the coach's training objectives	Rebuild fitness	Rebuild fitness Friendly games to test players and tactics of the team	Pretournament Taper Training Nutrition plan and psychological strategies for the next tournament	Beginning of the tournament

Figure 9.6 Suggested activities for each microcycle between national league and international tournament

As visible in the above figures, any competition plan is a complex and highly integrated activity. In order to achieve your competition objectives, you have to always create a plan where you have to consider:

1. Rest, recover/regenerate to remove the postgame fatigue
2. Rebuild fitness to avoid detraining prior to the beginning of the next tournament
3. Rebuild fitness and participate in friendly games to build team cohesion

4. Pretournament taper: use training, nutrition and psychological techniques to facilitate supercompensation for the tournament

5. Participate in the planned international tournament

NUTRITIONAL CONSIDERATIONS FOR TAPERING AND PEAKING

Although there has been an abundance of studies looking at the effects of different feeding protocols prior to competition, very few have explicitly looked at nutritional strategies for peaking and tapering into the competition. For endurance sports, the use of concepts like carbohydrate loading has been well supported and widely accepted in both science and practice, and even more recently strategies to promote a more fat adaptive state have been explored (Jeukendrup & Gleeson, 2010). However, at the current time, strength and power sports lack major evidential support in the literature. This is not because the studies conducted were poor, but rather there is a limited number of studies and most of them are focused on acute manipulations rather than the context of systematically of peaking and tapering for competition. This section seeks to outline fundamental nutritional recommendations for athletes during peaking and tapering phases based on somewhat limited direct evidence, and an abundance of strong theoretical evidence that can be universally applied.

The world of sports nutrition, similar to sport and exercise science, is often subject to fads, trends, and pseudoscience. It can be difficult to discern what areas of nutrition to focus on based on their relative effect size on performance and also how much empirical and anecdotal evidence they carry. How much of this supplement should we use, how much weight should I gain, what type of fruit should I be eating, how many grams of this micronutrient should I be taking in? These are all valid questions, but what are the effects of such manipulations, and are there other areas which warrant more powerful investigation? Although science is fluid and concepts may come and go as time goes on, at the current time it is suggested that athletes focus on the following nutritional considerations during peaking and tapering phases:

- Following previously established eating patterns.
- Calorie balance and maintaining bodyweight.
- Substrate repletion by prioritizing carbohydrate intake.
- Hydration and body water manipulations.

Let us take a look and see how these considerations should be used in an effective peaking and tapering program.

FOLLOWING ESTABLISHED EATING PATTERNS

Contrary to popular belief, the peaking and tapering phases are not the time to develop new eating habits or routines with an athlete. This will likely provide a major distraction from the training process, and unfortunately tapering is generally an insufficient amount of time to significantly benefit from major nutritional interventions. Major nutritional interventions and patterns should be introduced early in the training cycle, such as the general preparatory phase, so that the athlete has time to practice acclimatizing to their new routine before any major competitions. Once the athletes have entered the specific preparatory phase, these habits should be well ingrained and practiced, thus proceeding into their competitive periods, making only minor adjustments as needed.

Imagine you are coaching a group of gymnasts prepping for the world championships. Two weeks before they compete, once they have begun to master their routines, you ask them to add one incredibly technical and difficult move to each event for which they have minimal experience with. Some may catch on and adapt, while the majority will likely have difficulty and spend valuable time and recoverability working on these new moves instead of mastering their existing routines. A similar thought can be applied to nutritional interventions at this time. Even if the established routines are not perfected yet or potentially suboptimal in some ways, it is likely better to only make small modifications at this point, as any major change can be a significant potential distraction from the training process. Coaches, athletes, and the sport science team can seek to improve the strategies in subsequent training cycles after the major competitions have been completed.

It should be stressed to the coaches and athletes not to begin protocols which are new or ones which have not been familiarized to the athlete yet. For sports in which macronutrient manipulations are commonly made in the tapering phase, such as carbing up or down, athletes should rehearse and practice these protocols during less significant or important competitions, so they have a feel and understanding of them for the competitions for which they are peaking (Benardot, 2006; Burke, 2007; Jeukendrup & Gleeson, 2010). For weight class sports where body water manipulations are commonly used, the athletes should rehearse making weight under these conditions in earlier competitions, so they know how it will feel and the best strategies to subsequently recover any lost water. This also includes arrangements for travel, where the availability of certain foods or resources might not always be the same. Coaches and athletes should do their homework and see if they will be able to maintain such patterns while away or take precautionary steps in prepping food prior to traveling. There should be no surprises with a nutritional strategy leading into the competition.

CALORIE BALANCE AND MAINTAINING BODYWEIGHT

In many sports, particularly weight class sports such as combat sports, powerlifting, and weightlifting, it is commonplace for athletes to continue altering their bodyweight all the way up until the day of the competition. Altering bodyweight during the peaking and tapering phases, although may seem like a necessary evil, truly goes against what this phase seeks to accomplish. Now this may seem like an exaggeration, but think about it, what does peaking represent? For that training cycle, it represents a culmination of perfection in the athlete's sport skills and tactics, fitness characteristics, and psychological readiness. Ironically, altering one's bodyweight can potentially adversely affect all these aforementioned areas.

Sport skills and techniques are constantly being refined through feedforward and feedback mechanisms. During the peaking and tapering phases, particularly for more advanced athletes, it is essential that the athlete maintains a "feel" or familiarity with their skills and techniques. The feel for these techniques can be altered with significant changes in bodyweight, sometimes for better or worse, but the inconsistency of established feel can potentially be problematic. For example, a soccer player's ability to chase down a pass and resist a slide tackle, a rugby player's ability to maintain her posture in the scrum, or a powerlifter's ability to abdominally brace against his weightlifting belt during a maximal squat—all these techniques or abilities may be slightly altered by changes in bodyweight.

Similarly, changes in bodyweight may have potential effects on fitness characteristics. Of particular note are strength, power-to-weight ratio, metabolic efficiency, and others. Again, some of these changes could potentially be positive or negative; however, it is the inconsistency in performance that coaches should seek to avoid. Athletes should feel very comfortable with how fast they can move, how high they can jump, and their ability to express strength in a number of different scenarios within the peaking and tapering phase and should not be looking to correct or adjust more than necessary. It is also important to note that this training period is deliberately designed to help alleviate accumulated fatigue and allow dormant fitness characteristics to be expressed to their maximal capacity. The weeks leading into the competition already serve to calibrate these changes in fitness and express them into improved performance. Adding an additional layer of difficulty in this process from major bodyweight changes further delays this process and may prevent the athlete from getting the feel they need to compete. Additionally, the fatigue of hypocaloric dieting has been well documented to produce performance decrements across the entire fitness spectrum (Loucks, 2004). Major calorie restriction leading into a competition will have a direct negative effect in the ability to express technique, speed, power, and force, which will be a major hindrance during the time when the athlete should generally be experiencing the best training of that cycle.

One of the last issues on the athlete's mind during the peaking and tapering phase are strategies to continue changing bodyweight. As previously mentioned, acute manipulations should already be practiced and well rehearsed, but the anxiety of being too heavy or too light for the competition is an unwelcome distraction. Additionally, if the athlete is struggling to continue manipulating bodyweight leading into the competition, they will certainly not be focused on the competition itself as much as they could be, and they will potentially seek out other training and nutritional methods themselves to obtain the bodyweight, such as eating to discomfort and gastrointestinal distress, doing extra cardio work, starvation, or dehydrating. All these disparate methods are fatiguing in their own unique way, and instead of reducing fatigue, which is the theme of peaking and tapering, they continue adding fatigue and adversely affect performance.

With these thoughts in mind, it is generally recommended that athletes maintain an isocaloric state during the tapering and peaking phase. This should be started as early as the specific preparatory phase and continued until the end of the training cycle, or when all major competitions have ceased. Throughout this time the athlete's bodyweight should remain fairly stable, including normal daily fluctuations. For weight class-dependent sports, athletes should seek to be within 1-2% of their ideal competition body weight about 1-2 months prior to competition depending on their schedule when they enter the specific preparatory phase of their competitive cycle. Doing so allows the training to not be adversely influenced from efforts used to alter body composition while still allowing the athlete to make any last-minute weight modifications without having to resort to more drastic and potentially performance inhibiting approaches. Maintaining the isocaloric state can simply be achieved just through monitoring of caloric intake and athlete bodyweight.

PRIORITIZING CARBOHYDRATES

Carbohydrate is the primary energy substrate for virtually all sporting and exercise movements. Its role in performance enhancement has been well established in the literature and in practice for many years. Athletes who do not replenish lost carbohydrate or deliberately constrict carbohydrate virtually always show decreases in performance from baseline measures or from low/high carbohydrate group comparisons (Benardot, 2006; Burke, 2007; Burke, Cox, Culmmings, & Desbrow, 2001; Burke, Kiens, & Ivy, 2004; Cermak & van Loon, 2013; Gropper & Smith, 2012; Hausswirth & Mujika, 2013; Jeukendrup & Gleeson, 2010). Carbohydrate is also a vital component of setting the recovery-adaptive (RA) processes in motion following exercise (Burke et al., 2004; Cermak & van Loon, 2013; Hausswirth & Mujika, 2013; Howarth et al., 2010; Zoorob, Parrish, O'Hara, & Kalliny, 2013). Accordingly, replenishing endogenous carbohydrate stores is critical for peaking and tapering periods to ensure the athlete is performing at their maximal capacity and setting favorable cellular conditions for RA processes.

Previous research on the timing of carbohydrate intake and the sources of carbohydrate initially appeared extremely promising; however, more recent evidence seems to suggest that timing and type of carbohydrate intake may not have the effect we once thought (Aragon & Schoenfeld, 2013; Beelen, Burke, Gibala, & van Loon, 2010; Mondazzi & Arcelli, 2009; O'Reilly, Wong, & Chen, 2010; Zoorob et al., 2013). They may not be the most powerful considerations; however, they do appear to be meaningful and worthy of consideration when optimizing nutritional approaches. Recommendations generally revolve around prioritizing carbohydrate feedings around training times, specifically the pre-, intra-, and post-training periods. During peaking and tapering periods, it is generally recommended that athletes consume a carbohydrate-rich meal, one to three hours prior to training, which can consist of high or low glycemic carbohydrate sources depending on how proximal the meal is to the training session. Athletes should also consider consuming a liquid high glycemic carbohydrate during training sessions lasting generally more than about an hour, though this may not be necessary for sessions of shorter duration. Lastly, athletes should consume a carbohydrate-rich meal immediately following exercise, preferably from moderate to high glycemic sources at a rate of about 1 g per kilogram of bodyweight per hour in the hours following training sessions (Beelen et al., 2010).

Although the approaches adopted during peaking and tapering phases should closely mirror what was previously done in the specific preparatory phase, one difference that arises is a greater relative contribution of carbohydrates in total daily calories. Here because carbohydrate is so tightly linked to performance and recovery, calories from other sources can be slightly reduced to ensure optimal performance conditions for the athlete. In order to maintain muscle mass, athletes generally should not reduce daily protein intake to levels less than about 1.3 g per kilogram of bodyweight per day (Leidy et al., 2015; Westerterp-Plantenga, Nieuwenhuizen, Tome, Soenen, & Westerterp, 2009). Generally, daily fat intake is more flexible than carbohydrate and protein in terms of effects on performance and in the short term can be reduced significantly with those calories being replaced by carbohydrates. As a guiding heuristic, athletes should not reduce fat to less than about 0.66 g per kilogram of bodyweight per day for extended periods of time due to the potential for unfavorable endocrine responses and issues with joint health, but since peaking and tapering periods are generally short in practice, this reduction in fat to prioritize carbohydrate appears favorable mostly in the short-term (Bardner & Shoback, 2011; Borer, 2013; Hamalainen, Adlercreutz, Puska, & Pietinen, 1984; Hamalainen, Adlercreutz, Puska, & Pietinen, 1983). Protocols seeking to promote a more fat-adapted state for the athlete have shown positive effects in some physiological variables; however, often this still comes at the expense of exercise performance. Although future research on fat-adapted protocols may shed more light on its use and application, at the current time it seems difficult to argue against high carbohydrate protocols given the plethora of support.

HYDRATION AND BODY WATER MANIPULATIONS

Hydration is critical throughout the training process, and this is no different during peaking and tapering periods. Even light reductions in bodyweight (1-2%) from fluid loss are associated with decreased performance and continue to increase in severity ranging from severe cramping (2-5%) all the way up to changes in heart rate and hallucinations (7-10%) (Benardot, 2006; Casa et al., 2000; Hausswirth & Mujika, 2013; Jeukendrup & Gleeson, 2010; Judelson et al., 2007; Maughan & Shirreffs, 2010). Needless to say, adequately hydrating and replacing lost fluids and electrolytes is essential during the peaking and tapering processes. Athletes with a high training age generally can autoregulate thirst and hydration well, though younger athletes or athletes training in hot environmental conditions may need to be more conscious of hydrating. General recommendations include consuming water pre- and post-exercise, consuming about 1.5 times the amount of water lost until voluminous and fairly clear urination resumes. This may need to be adjusted based on the volume and intensity of the exercise and the exercise conditions. Endurance athletes training and competing in the heat should also consider replacing lost electrolytes for the prevention of hyponatremia.

For weight class sports, athletes seeking to maximize their lean body mass at their weight class often resort to water cuts in order to lose the last remaining portion of the weight. The water cut can be an effective strategy during peaking and tapering phases to make competition weight; however, it is limited by the severity of weight cut needed and the time allotted for rehydration. Some sports such as Brazilian Jiu-Jitsu often utilize mat-side weigh-ins, allowing virtually no time to rehydrate, whereas others may utilize 24-hour weigh-ins, allowing for greater time to rehydrate. Generally, a well-hydrated athlete can lose upwards of 5% of bodyweight without any major health complications. However, it is not recommended for athletes to cut ≥ 7% of bodyweight due to the increased health risks. Keep in mind, this does not imply that they will be without performance decrements, and complete rehydration may take upwards of 24 hours to occur.

Basic water-cutting protocols generally involve reducing water, carbohydrate, and sodium during the peaking and tapering phase. Because athletes store quite a bit of carbohydrate in the form of glycogen, a very low carbohydrate diet is typically adopted upwards of five days prior to and all the way up until the weigh-ins. Sodium, like carbohydrate, tends to pull water into the cells. Large reductions in sodium intake generally occur within two days prior to and all the way up until the weigh-ins. Last but not least, reductions in actual water intake generally only need to occur 24 hours prior to and all the way up to the weigh-ins. Some athletes choose to begin cutting water weeks prior to competing, and unfortunately, this does not provide any additional benefit but results in the potentially massive cost of prolonging dehydration. Advanced protocols can also involve water loading the week before to enhance water loss; later on, however, this should only be done under the close supervision of an experienced coach and with medical approval. All water-cutting protocols should be done with physician approval, as they are all potentially dangerous. Following the water-cut and weigh-in, the athlete should seek to rehydrate using half parts plain water and carbohydrate and sodium-rich foods such as sports drinks, juice, kids cereal, frozen yogurt, soda, and so on.

PSYCHOLOGICAL SUPPORT FOR TAPERING AND PEAKING

In the competitive phase of most individual and team sports, the athlete or the team is required to reach a number of peaks. An athlete in an individual sport may take part in more than 10–20 competitive events during a season, and team sport athletes may play more than 40 games (e.g., soccer, handball) or even more than 70 games (e.g., basketball) during a season. This information requires special attention, since during a season an athlete participates in a major target competition (e.g., European/World Championships, NBA playoffs) and subtarget competitions (e.g., league games/tournaments). During the tapering and peaking times for competition, an athlete is expected to perform at his/her best, having to "pull it all together" on various occasions over a period of a few months, avoid injuries, and maintain a high level of motivation, self-confidence, and attention-focusing with a high level of self-regulation.

The duration of tapering and peaking for a target competition may take 7–10 days (Bompa & Haff, 2009). During that time the integration between all the athlete's preparations (i.e., physical, psychological, technical, and tactical) should be mutually efficient and precise. An important point is that throughout the 7–10 days, the physical load in regards to volume, intensity, and duration decreases, and on the other hand, the psychological tension increases. Major reasons for the increase of the psychological tension might be the change in the training routine and load, uncertainty regarding competition conditions/results, and media attention. Other stressors might be sleeping problems, negative thoughts, life stressors (e.g., family, financial, relationship), conflicts with coach/team member/official/media, travel concerns (e.g., length of flight, jet lag adjustments), and local conditions in training/competition (e.g., weather, equipment).

Psychological support during these 7–10 days can be divided to four stages. The first stage is the **habituation** stage, which lasts for a few days depending on the length of the flight. During this stage the athlete has to adapt to the training site and local conditions, such as weather, time zone, and food. The psychological support in this stage includes practice and perfecting the main psychological skills that are needed for peak performance. Examples might be relaxation, imagery (i.e., the technical/tactical side of performance), concentration, and self-talk. It is recommended to support the psychological practice with biofeedback techniques for strengthening the athlete's self-confidence. The framework for conducting the psychological practice might be a hotel room or training location, while the focus is on the ability of the athlete to practice the psychological skills/techniques by himself/herself.

The second stage is the **psychological routine**, which is carried out up to two days before the competition. The objective of this stage is to help the athlete return to his/her ordinary training routines, from the physical and psychological perspectives. In this stage individual psychological sessions are conducted, not only in a consultation room but also at the training facilities, before and after practice. This strengthens the connection between psychological techniques and performance and improves the ability of the athlete to self-regulate and self-control himself/herself during the days before competition. During this stage the evaluation of psychological variables for peak performance is essential. This can be done by performing a package of psychological exercises and by achieving specific results, which are linked to the athlete's psychological readiness in different sports (see figure 9.7).

	Judo	Tackwondo	Rhythmic Gymnastics	Swimming	Rowing
RTP (ms) 10 Simple 20 Choice 20 Discrimination	115–145, ratio: 9:1 140–168, ratio: 8:2 127–136, ratio: 9:1	127–130, ratio: 8:2 140–164, ratio: 8:2 122–138, ratio: 9:1	N/A	N/A	N/A
STEP: Stop reaction exercise (SRE)	0.6–0.7 sec	0.6–0.8 sec	N/A	0.11–0.13 sec	0.13–0.14 sec
Time reproduction exercise (TRE)	4.94–5.06 (±0.06)	4.95–5.05 (±0.05)	4.95–5.05 (±0.05)	4.97–5.03 (±0.03)	4.94–5.06 (±0.06)
Δ GSR (kΩ) (30 sec, 1 min)	Δ 600–700 (1 min)	Δ 600–650 (1 min)	Δ 450–500 (30 sec)	Δ 500–600 (30 sec)	Δ 500–600 (30 sec)
Performance time during imagery	N/A	N/A	Individual: 1.29–1.30 Group: 2.29–2.30	Approximately the personal best time	Approximately the personal best time

Figure 9.7 Examples of measurements for evaluating psychological readiness associated with peak performance

The procedure of using response training program (RTP) exercise is explained in Chapter 6. Regarding the table's results, it is important to pay attention to the ratio data. In the simple mode, there are 10 attempts (9 fast and 1 slow) in the Judo example. However, in the choice and discrimination modes, the attention is only on the dominant side of the athlete, and therefore the ratio is based on only 10 attempts and not 20.

The simulation training exercise program (STEP) includes two motor exercise: (a) stop reaction exercise (SRE) and (b) time reproduction exercise (TRE). In SRE the athlete held a wide stopwatch while his/her thumb press on the start-stop button. The main goal of this exercise is to start and stop the time as fast as possible. To achieve good results, the athlete must learn how to relax his/her muscles and concentrate on the act itself while finding the optimal balance between muscle tension and the necessary concentration level. In TRE the athlete is asked to reproduce a time period of 5 seconds without looking at the time clock. After 3–4 training attempts, the athlete has 5 attempts. The goal is to reproduce a time that is close to 5 seconds based on sport discipline (see figure 9.7).

Another measurement is the Galvanic Skin Response (GSR) (kΩ). In this biofeedback exercise, athletes are asked to relax and concentrate for a specific time period (e.g., 30 sec in combat sport and 60 sec in other sports). The delta achieved indicates the ability of the athlete to self-regulate. Finally, the last exercise is performance time during imagery. The goal is to imagine a clean performance within the ideal competitive performance time. For example, the personal best time in 100 m swimming or performing a rhythmic gymnastics routine within the required competition time limit (e.g., 1:30 min). In some cases, both the coach and athlete should make the necessary corrections in the athlete's preparation in order to compensate for not achieving the above measures. For example, measures of RTP and STEP based on Figure 9.7 inform athlete about his/her speed decision making, concentration, and the ability to create an optimal balance between concentration and relaxation. In cases where the athlete demonstrates an insufficient result in these exercises, he or she must pay more attention to the technical details of his/her performance (especially with a ratio of 7:3 or 6:4 in RTP). Accordingly, corrections should be made in the warm-up preparation and tactical preparation for the upcoming fight in combat sports. In addition, throughout the competition process, an athlete may achieve his/her best results in the RTP and STEP exercises. For example, after a first victory in judo and in the preparation process for the second fight, the athlete may achieve 0.6 sec in SRE, and 4.97, 5.01, and 4.96 in the three attempts of the TRE.

The data in figure 9.7 are based on more than 40 years of experience working with elite athletes from a variety of sport disciplines and countries (Blumenstein & Orbach, 2012a,b). During the individual sessions in the psychological routine stage, the athlete practices the protocols of the PST described in Chapters 3 and 6. Psychological preparation for the competition includes the LMA approach, RTP, and STEP. For example, in combat sport the athlete uses a focusing-attention technique in order to perfect the ability to effectively ignore external distractions and an imagery technique to review some of his/her new drills that have been prepared specifically for competition. Imagery (i.e., imagining **only** successful performance) is provided for a time period specified to the length of the combat (e.g. in judo 4 minutes, taekwondo 2 minutes × 3 times). RTP is used to prepare the athlete to stay calm and relaxed between rounds, matches, and fights. In addition, every evening when athlete is alone in his/her room, he or she is asked to practice imagery and self-talk and to remember what he or she has to do in his/her upcoming sessions. Finally, 10-minute muscle relaxation is performed.

The third stage, **specific psychological preparation**, usually takes place during the last two days before competition (including on competition day). This stage is focused on preparing the athlete for his/her major competitive events, while emphasizing the first part of the competition (e.g., preliminaries, heats, qualification round, first game). Psychological techniques are practiced within time limitations, mimicking the competitive event specifically to the given sport, and focused on developing a competition plan to be used in the initial part of the competition. Self-talk (e.g., "I'm ready," "I'm calm and focusing on the technical performance," "I can do it"), imagery performed when competing against his/her first opponent, positive thinking (e.g., "remember: thought makes motion"), short RTP and STEP, and relaxation with GSR biofeedback, are often performed in combination in order to strengthen his/her self-confidence and psychological readiness for competition. It is interesting to note that accurate objective measures such as RTP, STEP, and GSR biofeedback, which inform the athlete about his/her self-regulation, concentration, and readiness, have a positive effect on the athlete's optimal psychological state. During the last 24 hours before competition, the athlete performs a *precompetitive activity routine* (PCA-R), which includes special procedures concerning actions, thoughts, and rituals. PCA-R helps athletes achieve an optimal feeling of mental readiness, regulates emotions, and increases self-confidence. PCA-R is based on the athlete's retrospective positive performance experiences, from which he or she selects different elements of precompetitive activity, actions, thoughts, and psychological strategies that correlate with successful results in the past. Examples might include relaxation, self-talk, imagery, habitual clothes, sitting in a special place on the bus, food, length and content of warm-up, and various routines and rituals to be performed 24 hours before competition. In team sports, the athlete relates his/her personal routine with the structure and schedule of the team preparation, for example, team meetings, small group meetings (e.g., defense, offense), talking with the coach, and team warm-up.

A short time before the actual performance in competition, the athlete can use a *preperformance routine* (PPR) in order to create an optimal body-mind state for peak performance (for more details, see chapter 3). PPR is a significant part of athletic performance, during which athlete utilizes his/her plan of behavior, thoughts, and feelings, to prepare for the performance. In general, PPR is a ritual scenario of what an athlete/team does, thinks, and feels in order to get into the "zone" prior to the actual performance. There are many examples of PPR in different sports that are described in the psychological literature (e.g., Blumenstein & Orbach, 2012b; Lidor, 2007). For example, PPR for a free throw in basketball includes holding the ball, self-talking, dribbling, and focusing attention on the basket before the actual throw. PPR for a penalty shot in soccer includes having a positive attitude, deep respiration, observation of the goal keeper position, making a decision regarding the execution of the penalty shot, visualization of a successful penalty shot, and execution of the penalty shot. PPR for 100 m swimming sprint includes positive self-talk, visualization of the sprint, deep respiration together with muscle relaxation, anticipation of an explosive start, reacting to start ritual commands, and focusing on the start movement and the signal.

The athlete demonstrates his/her PPR automatically, quickly, and precisely. All psychological techniques that are applied in this time period must be short (e.g., relaxation 5–10 sec), punctual and combined at exactly the right time, with full concentration and high confidence. There are times in competition, especially in athletics, that athlete asks for applause-support from spectators at the stadium to optimize/excite his/her state before the attempt/jump/throw. In another word, the use of the applause changes his/her PPR, and this was not practiced before competition. The athlete is excited by the applause, but may not be able to fully concentrate on performance technique.

In addition, the athlete has no control over the applause rhythm, which may not correlate with his/her performance rhythm. Therefore, on some occasions only athletes with a high self-regulation level and stable technique can use this support. Finally, PPR has individual peculiarities that are linked with the athlete's experience and with the competition demands.

Lastly, *postperformance activity* (PPA) takes place following the performance, and includes a postperformance analysis and actions in order to prepare the athlete for the next attempt, which might start a few minutes or a few hours after the previous attempt (e.g., fights in combat sport, jumps in athletics). In this period, the athlete's first self-estimates as to what has just happened emphasizes positive elements for upcoming performance and in some cases makes a conclusion as to what needs to be improved in the next attempt/fight/throw. During the time before PPR for the upcoming performance, the athlete plans in advance what kind of actions and thoughts to engage in so that he or she can conserve emotional and physical energy. In most of the cases athlete engages in activities such as muscle relaxation, breathing, imagery, or just walking and listening to music. When the time of the upcoming competition arrives, the athlete begins a short warm-up and his/her regular PPR.

The fourth stage is **recovery**, which is provided one or two days after competition. The objective of this stage is to help the athlete start recovering from the extreme physical and psychological efforts he or she has made during the competition. In this stage, one or two 30-minute psychological recovery sessions are conducted with each athlete. The session includes an analysis of the positive and negative sides of competition's results, and relaxation techniques while listening to music. After returning home from the competition, it is recommended that a few individual/team sessions be conducted regarding future cooperation with a sport psychology consultant and new goals for the next competition.

DIFFERENT ROUTINES AND RECOVERY TECHNIQUES IN INDIVIDUAL AND TEAM SPORTS

In this section, different *PCA-R /PPR and recovery techniques* for individual and team sports, based on several competition schedules, will be discussed.

1. *One-day competition plan for individual sport*: For example, in combat sports (e.g., Olympic content) the competitive day includes an elimination round of 64, of 32, of 16, then quarterfinals, a rest for several hours, and then repechage semifinals and final combats. The athlete prepares for the first fight, and after victory he or she has approximately 40 minutes for recovering energy and developing a tactical plan in order to get ready for the next combat. In addition, the athlete's goal is to attain a high level of concentration and self-confidence for the beginning of the next combat. In this time period the athlete disconnect mentally from the previous combat, replaces energy with psychotherapeutic techniques (e.g. message), performs muscle relaxation, and eventually develops a mental strategy and tactical plan for the next combat. Then, there is the physical warm-up, with his/her regular PPR being applied. The rest

pause before the repechage, semifinals, and final combat may last for a few hours. During this time, the athlete usually stays in the warm-up hall/area or finds a calm room for sleep and relaxation (sometimes with listening to his/her favorite music), eats something light, has a relaxation massage, and/or engages in deep muscle relaxation (i.e., 20–25 min). It is important to avoid talking about the combat and to connect only with close athlete's staff. Approximately 1 hour before the combat, the athlete begins his/her regular physical warm-up and creates a positive attitude with self-talk ("I feel good," "I am full of energy," "I have a good feeling"). Specific judo exercises and light workouts with a sparring partner should be performed in an optimistic and enthusiastic atmosphere and should focus on specific judo preparation in order to increase the athlete's concentration and self-confidence. Then, the athlete and coach should discuss and practice main tactical moments of the upcoming combat followed by short local relaxation and visualization of a successful performance. Finally, the athlete prepares to exit to the combat with self-confidence, focus, and clear tactical plan.

In athletics, during the qualifying round, the athlete has three attempts in long/triple jump and throws. The athlete uses his/her regular PPR before the first attempt and a short version of PPR between attempts, which includes a short warm-up together with local relaxation, positive self-talk ("I have an additional chance, stay calm," "I can do it now"), visualization of ideal techniques for the performance, and finally attention focus. In addition, between attempts, PPA as described above is applied.

2. *One day competition plan for team sport*: In team sport PCA-R and PPR include individual and team routines, which are influenced by numerous cultural, local/club, and social differences. However, there are some obligatory components that are important for individual/team PCA-R: (1) a good night's **sleep** before the game (approximately 8 hours) plus a 1–2 hour nap in the afternoon and preparing equipment in his/her bag in the previous evening; (2) **visualizing** fragments of the game (i.e., a good performance) the night before, in order to train the player's mind for a situation that might take place on the following day; (3) the usual **meals**; (4) **positive thoughts and a calm state** (sometimes with listening to favorite music using iPods or watching movies); and (5) **positive and strong team cohesion** based on "one team, one spirit." In PCA-R for team sport, special attention should be given to team cohesion. This can be done by instructing the experienced players to connect with the young players and to take responsibility for strengthening team cooperation. In addition, the coach plays a dominant role in this period, conducting individual and team meetings and motivating athletes before they get on the field.

   PPR before a soccer game also includes individual/team meetings with the coach, travel by bus from the hotel, and getting ready in the dressing room. In addition, there is a team warm-up that includes standard exercises and a kick-off up to 10–15 minutes before the game start. Then, the players are back in the dressing room—they put their shirt and shin pads on and then go out to the field for the start of the game. The main results of PCA-R and PPR are to help players to create a confident, focused, and relaxed mind-set, which helps them to play freely and trust in their skills.

3. *Two- to three-day competition plan:* It consists of a qualifying round on one day and then a pause for one or more days, followed by semifinals and the final competition. The athlete uses one regular PCA-R and

PPR for achieving qualifying criteria, investing minimum psychological and physical efforts. Then, after achieving the criteria for the semifinals/final, the athletes engage in relaxation and recovery activities for better preparing for the competition. Relaxation techniques are applied by the athlete for 25–30 minutes in his/her room each day in the afternoon after practice. In addition, the athlete completes his/her day with relaxing activities such as reading book, watching movies, playing a computer game, and engaging in light physical activity. The main objectives of this period are to maintain optimism and high professional level and to avoid negative thoughts about the upcoming competition. Each day includes imagery (10–15 min) of ideal performance techniques during practice and in the athlete's room (not before sleep).

4. *PCA-R in combat sports for rapid weight loss:* Combat athletes are classified according to their body mass. Therefore, many athletes acutely reduce their body mass in an attempt to gain an advantage by competing against lighter, smaller and weaker opponents. In most combat sports (e.g., wresting, judo, taekwondo), rapid weight loss is prevalent and can be part of PCA-R. Most athletes reported that their greatest bodyweight reduction was approximately 5% of bodyweight in the week preceding the weigh-in. Usually the weight loss is achieved by a variety of methods, such as the use of a sauna, reducing energy intake, and fasting one day prior to the weigh-in. It is important to note that rapid weight loss is associated with significantly higher scores on anger, fatigue and tension, and lower scores on vigor (Hall & Lane, 2001). Psychological support in PCA-R (usually fasting one day) includes goal setting and sport motivation, practicing self-discipline and self-control, muscle relaxation, and focusing on behavioral changes with modified thinking about himself/herself. It is recommended in this period to engage in relaxing activities such as listening to favorite music, reading, light conversation (i.e., avoiding talk about food), and playing computer games.

Based on the above, the following remarks summarize the major considerations regarding psychological support for tapering and peaking for competition. The coach-athlete relationship is the foundation for the successful application of the above remarks in sport. The coach plays the dominant role in PCA-R (especially in team sports) and in PPR. The sport psychology consultant practices and ensures that the athlete will perfect his/her psychological skills, develops a competition behavioral model for the athlete and coach, and, finally, corrects possible mistakes. However, on the last day before an actual performance, the coach is the main figure and is the closest figure to his/her athletes. Eventually the coach and athlete are the ones who use the above psychological program in the actual performance. The sport psychology consultant provides support and performs consulting work with the athlete and coach. Therefore, full cooperation between athlete-coach-sport psychology consultant is an important factor for success and peak performance.

REFERENCES

Acharya, J., & Morris, T. (2014). Psyching up and psyching down. In A. G. Papaioannou & D. Hackfort (Eds.), *Routledge companion to sport and exercise psychology: Global perspectives and fundamental concepts* (pp. 386–401). Hove, East Sussex, UK: Routledge.

Achten, J., & Jeukendrup, A. E. (2003). Applications and limitations. *Sports Medicine, 33*(7), 517–538.

Amann, M. (2011). Central and peripheral fatigue: Interaction during cycling exercise in humans. *Medicine & Science in Sports & Exercise, 43*(11), 2039–2045.

Anshel, M. H., & Payne, J. M. (2006). Application of sport psychology for optimal performance in martial arts. In J. Dosil (Ed.), *The sport psychologist's handbook: A guide for sport-specific performance enhancement* (pp. 353-74). Inglaterra: John Wiley & Sons.

Antonini-Philippe, R., Reynes, E., & Bruant, G. (2003). Cognitive strategy and ability in endurance activities. *Perceptual and Motor Skills, 96*(2), 510–516.

Aragon, A. A., & Schoenfeld, B. J. (2013). Nutrient timing revisited: Is there a post-exercise anabolic window? *Journal of the International Society of Sports Nutrition, 10*(1), 5.

Balague, G. (2000). Periodization of psychological skills training. *Journal of Science and Medicine in Sport, 3*(3), 230–237.

Banister, E. W., Carter, J. B., & Zarkardas, P. C. (1999). Training theory and taper: Validation in triathlon athletes. *European Journal of Applied Physiology, 79*, 182–191.

Bar-Eli, M., & Blumenstein, B. (2004). The effect of extra-curricular training with biofeedback on short running performance of adolescent physical education pupils. *European Physical Education Review, 10*(2), 123–134.

Baxter-Jones, G., & Maffulli, N. (2003). Endurance in young athletes: It can be trained. *British Journal of Sports Medicine, 37*, 96–97.

Beauchamp, M. K., Harvey, R. H., & Beauchamp, P. H. (2012). An integrated biofeedback and psychological skills training program for Canada's Olympic short-track speed skating team. *Journal of Clinical Sport Psychology, 6*, 67–84.

Beelen, M, Burke, L. M., Gibala, M. J., & van Loon, L. J. C. (2010). Nutritional strategies topromote postexercise recovery. *International Journal of Sport Nutrition and Exercise Metabolism, 20*(6), 515–532.

Behringer, M., Vom Heed, A., Yue, Z., & Mester, J. (2010). Effects of resistance training in children and adolescents: A meta-analysis. *Pediatrics, 126*(5), e1199–e1210.

Bellisle, F., McDevitt, R., & Prentice, A. M. (1997). Meal frequency and energy balance. *British Journal of Nutrition, 77*, S57-70 (suppl. 1).

Benardot, D. (Ed.). (2011). *Advanced sports nutrition*. Champaign, IL: Human Kinetics.

Blumenstein, B., Bar-Eli, M., & Tenenbaum, G. (1995). The augmenting role of biofeedback: Effects of autogenic, imagery, and music training on physiological indices and athletic performance. *Journal of Sports Sciences, 13*, 343–354.

Blumenstein, B., Bar-Eli, M., & Tenenbaum, G. (1997). A five-step approach to mental training incorporating biofeedback. *Sport Psychologist, 11*, 440–453.

Blumenstein, B., Bar-Eli, M., & Tenenbaum, G. (Eds.). (2002). *Brain and body in sport and exercise: Biofeedback application and performance enhancement*. Hoboken, NJ: Wiley.

Blumenstein, B., & Lidor, R. (2004). Psychological preparation in elite canoeing and kayaking sport programs: Periodization and planning. *Applied Research in Coaching and Athletics Annual, 19*, 24–34.

Blumenstein, B., Lidor, R., & Tenenbaum, G. (2005). Periodization and planning of psychological preparation in elite combat sport programs: The case of judo. *International Journal of Sport and Exercise Psychology, 3*, 7–25.

Blumenstein, B., Lidor, R., & Tenenbaum, G. (Eds.). (2007). *Psychology of sport training*. Oxford, UK: Meyer & Meyer Sport.

Blumenstein, B., & Orbach, I. (2012a). *Mental practice in sport: Twenty case studies*. Hauppauge, NY: Nova Science Publishers.

Blumenstein, B., & Orbach, I. (2012b). *Psychological skills in sport: Training and application*. Hauppauge, NY: Nova Science Publishers.

Blumenstein, B., & Orbach, I. (2014). *Biofeedback for sport and performance enhancement*. Oxford Handbooks Online. New York, NY: Oxford University Press. doi:10.1093/oxfordhb/9780199935291.013.001

Blumenstein, B., & Orbach, I. (2015). Psychological preparation for Paralympic athletes: A preliminary study. *Adapted Physical Activity Quarterly, 32*, 241–255.

Blumenstein, B., & Orbach, I. (2018). Periodization of psychological preparation within the training process. *International Journal of Sport and Exercise Psychology*. doi:10.1080/1612197X.2018.1478772

Boksem, M. A., Meijman, T. F., & Lorist, M. M. (2005). Effects of mental fatigue on attention: An ERP study. *Cognitive Brain Research, 25*, 107–116.

Bompa, T. (1956). *Antrenamentul in perioada pregatitoare* (Training Methods during the Preparatory Phase) (Vol. 3, pp. 22–28). Bucuresti: Caiet pentru sporturi nautice.

Bompa, T. (1999). *Periodization: The theory and methodology of training* (4th ed.). Champaign, IL: Human Kinetics.

Bompa, T., & Buzzichelli, C. (2015). *Periodization of training* (3rd ed.). Champaign, IL: Human Kinetics.

Bompa, T., & Haff, G. (2009). *Periodization: Theory and methodology of training* (5th ed.). Champaign, IL: Human Kinetics.

Bosquet. L., Monpetit, J., Arvisals, D., & Mujika, I. (2007). Effects of tapering on performance: A meta-analysis. *Medicine & Science in Sports & Exercise, 39*, 1358–1365.

Boutcher, S. H. (1992). Attention and athletic performance: An integrated approach. In T. Horn (Ed.), *Advances in sport psychology* (pp. 251–265). Champaign, IL: Human Kinetics.

Bradely, W. J., Cavanagh, B., Douglas, W., Donovan, T. F., Twist, C., Morton, J. P., & Close, G. L. (2015). Energy intake and expenditure assessed 'in-season' in an elite European rugby union squad. *European Journal of Sport Science, 15*(6), 469-79.

Bray, M. S., Hagberg, J. M., Perusse, L., Rankinen, T., Roth, S. M., Wolfarth, B., & Bouchard, C. (2009). The human gene mapfor performance and health-related fitness phenotypes: The 2006–2007 update. *Medicine & Science in Sports & Exercise, 41*(1), 35–73.

Burke, K. L., & Brown, D. (2003). *Sport psychology library series: Basketball.* Morgantown, WV: Fitness Information Technology.

Burke, L. M. (Ed.). (2007). *Practical sports nutrition.* Champaign, IL: Human Kinetics.

Burke, L. M., Cox, G. R., Culmmings, N. K., & Desbrow, B. (2001). Guidelines for daily carbohydrate intake: Do athletes achieve them? *Sports Medicine, 4*, 267–299.

Burke, L. M., Kiens, B., & Ivy, J. L. (2004). Carbohydrates and fat for training and recovery. *Journal of Sports Science, 22*(1), 15–30.

Burton, D., Naylor, S., & Holliday, B. (2001). Goal-setting in sport: Investigating the goal effectiveness paradigm. In R. Singer, H. Hausenblas, & C. Janelle (Eds.), *Handbook of sport psychology* (2nd ed., pp. 497–528). New York, NY: Wiley.

Cain, D. J., & Maffulli, N. (2005). Epidemiology of children's individual sports injuries: An important area of medicine and sports research. *Medicine and Sport Science, 48*, 1–7.

Calmels, C., Berthoumieus, C., & d'Arripe-Longueville, F. (2004). Effects of an imagery training program on selective attention of national softball players. *Sport Psychologist, 18*, 272–296.

Carrera, M., & Bompa, T. (2007). Theory and methodology of training: General perspectives. In B. Blumenstein, R. Lidor, & G. Tenenbaum (Eds.), *Psychology of sport training* (pp. 19–39). Oxford, UK: Meyer & Meyer Sport.

Casa, D. J., Armstrong, L. E., Hillman, S. K., Montain, S. J., Reiff, R. V., Rich, B. S., ... Stone, J. A. (2000). National athletic trainers' association position statement: Fluid replacement for athletes. *Journal of Athletic Training, 35*(2), 212–224.

Cermak, N. M., & Van Loon, L. J. (2013). The use of carbohydrates during exercise as an ergogenic aid. *Sports Medicine, 43*(11), 1139–1155.

Clayton, B., Brown, C., Burrell, L. M., & Matthews, M. D. (2011). The use of neuropeptide Y as a measurement of the effectiveness of stress inoculation. *West Point Resilience Project*, 1–16. www.dtic.mil/dtic/tr/fulltext/u2/a540922.pdf.

Collins, D. (1999). The psychology of power training. *Faster-Higher-Stronger, 3*, 28–29.

Collins, D., & MacPherson, A. (2007). Psychological factors of physical preparation. In B. Blumenstein, R. Lidor, & G. Tenenbaum (Eds.), *Psychology of sport training* (pp. 40–61). Oxford, UK: Meyer & Meyer Sport.

Cornier, M. A. (2011). Is your brain to blame for weight regain? *Physiology & Behavior, 104*(4), 608–612.

Cotterill, S. (2010). Pre-performance routines in sport: Current understanding and future directions. *International Review of Sport and Exercise Psychology, 3*(2), 132–153.

Croce, R. V. (1986). The effects of EMG biofeedback on strength acquisition. *Biofeedback and Self-Regulation, 11*, 299–310.

Csikszentmihalyi, M. (1990). *Flow: The psychology of optimal experience.* New York, NY: Harper & Row.

Cumming, F., & Williams, S. E. (2014). Imagery. In R. C. Eklund & G. Tenenbaum (Eds.), *Encyclopedia of sport and exercise psychology* (pp. 369–372). Thousand Oaks, CA: Sage Publications.

Davenne, D. (2009). Sleep of athletes—Problems and possible solutions. *Biological Rhythm Research, 40*(1), 45–52.

De Lorme, T., & Watkins, A. (1951). *Progressive resistance exercises.* New York, NY: Appleton-Century Croft.

Dosil, J. (Ed.). (2006). *The sport psychology handbook: A guide for sport specific performance enhancement.* Chichester, UK: Wiley.

Dotan, R., Mithell, C., Cohen, R., Klentrou, P., Gabriel, D., & Falk, B. (2012). Child-adult differences in muscle activation—A review. *Pediatric Exercise Science, 24*(1), 2–21.

Dressendorfer, R. H., Peterson, S. R., Moss Lovshin, S. E., Hannon J. L., Lee, S. F., & Bell, G. J. (2002). Performance enhancement with maintenance of resting immune status after intensified cycle training. *Clinical Journal of Sports Medicine, 12*(5), 301–307.

Eccles, D. W., & Riley, K. R. (2014). Psychological skills. In R. C. Eklund & G. Tenenbaum (Eds.), *Encyclopedia of sport and exercise psychology* (pp. 562–563). Thousand Oaks, CA: Sage Publications.

Elbe, A. M., & Kellmann, M. (2007). Recovery following training and competition. In B. Blumenstein, R. Lidor, & G. Tenenbaum (Eds.), *Psychology of sport training* (pp. 162–185). Oxford, UK: Meyer & Meyer Sport.

Elfhag, K., & Rössner, S. (2005). Who succeeds in maintaining weight loss? A conceptual review of factors associated with weight loss maintenance and weight regain. *Obesity Reviews, 6*(1), 67–85.

Enoka, R. M., & Duchateau, J. (2008). Muscle fatigue: What, why and how it influences muscle function. *The Journal of Physiology, 586*(1), 11–23.

Essig, K., Janelle, C., Borgo, F., & Koester, D. (2014). Attention and neurocognition. In A. G. Papaioannou & D. Hackfort (Eds.), *Routledge companion to sport and exercise psychology: Global perspectives and fundamental concepts* (pp. 253–271). Hove, East Sussex, UK: Routledge.

Evans, L., Jones, L., & Mullen, R. (2004). An imagery intervention during the competitive season with an elite rugby union player. *Sport Psychologist, 18*, 252–271.

Feltz, D., & Oncu, E. (2014). Self-confidence and self-efficacy. In A. G. Papaioannou & D. Hackfort (Eds.), *Routledge companion to sport and exercise psychology: Global perspectives and fundamental concepts* (pp. 417–429). Hove, East Sussex, UK: Routledge.

Fontani, G., Migliorini, S., Benocci, R., Facchini, A., Casini, M., & Corradeschi, F. (2007). Effect of mental imagery on the development of skilled motor actions. *Perceptual and Motor Skills, 105*, 803–826.

Forbes, G. B. (1987). Lean body mass-body fat interrelationships in humans. *Nutritional Reviews, 45*(8), 225–231.

Forbes, G. B. (2000). Body fat content influences the body composition response tonutrition and exercise. *Annals of the New York Academy of Sciences, 904*, 359–365.

Frey, M., Laguna, P., & Ravizza, K. (2003). Collegiate athletes' mental skills use and perceptions of success: An exploration of the practice and competition settings. *Journal of Applied Sport Psychology, 15*, 115–128.

Fry, R. W., Morton, A. R., & Keast, D. (1991). Overtraining in athletes: An update. *Sports Medicine, 12*(1), 32–65.

Gardner, D. G., & Shoback, D. (2011). *Greenspan's basic & clinical endocrinology* (9th ed.). New York, NY: McGraw-Hill Companies.

Goodger, K., Gorely, T., Lavallee, D., & Harwood, C. (2007). Burnout in sport: A systematic review. *The Sport Psychologist, 21*, 127–151.

Gould, D. (2002). The psychology of Olympic excellence and its development. *Psychology, 9*, 531–546.

Gould, D. (2015). Goal setting for peak performance. In J. M. Williams & V. Krane (Eds.), *Applied sport psychology: Personal growth to peak performance* (7th ed., pp. 188–206). New York, NY: McGraw-Hill.

Gould, D., & Carson, S. (2007). Psychological preparation in sport. In B. Blumenstein, R. Lidor, & G. Tenenbaum (Eds.), *Psychology of sport training* (pp. 115–136). Oxford, UK: Meyer & Meyer Sport.

Gould, D., & Damarjian, N. (1998). Mental skills training in sport. In B. Elliot (Ed.), *Applied sport science: Training in sport. International handbook of sport science* (Vol. 3, pp. 69–116). Sussex, UK: England: Wiley.

Gould, D., Flett, R., & Bean, E. (2009). Mental preparation for training and competition. In B. Brewer (Ed.), *Handbook of sport medicine and science: Sport psychology* (pp. 53–63). Chichester, UK: Wiley-Blackwell.

Gould, D., & Maynard, I. (2009). Psychological preparation for the Olympic Games. *Journal of Sport Sciences, 27*(13), 1393–1408.

Gould, D., & Udry, E. (1994). Psychological skills for enhancing performance: Arousal regulating strategies. *Medicine & Science in Sport & Exercise, 26*, 478–485.

Graf, S., Egert, S., & Heer, M. (2011). Effects of whey protein supplements on metabolism: Evidence from human intervention studies. *Current Opinion in Clinical Nutrition of Metabolic Care, 14*(6), 569–580.

Greenspan, M., & Feltz, D. (1989). Psychological interventions with elite athletes in competitive situations: A review. *Sport Psychologist, 3*, 219–236.

Hackfort, D., & Tenenbaum, G. (2014). Ethical issues in sport and exercise psychology. In A. G. Papaioannou & D. Hackfort (Eds.), *Routledge companion to sport and exercise psychology: Global perspectives and fundamental concepts* (pp. 976–987). Hove, East Sussex, UK: Routledge.

Hall, C. J., & Lane, A. M. (2001). Effects of rapid weight loss on mood and performance among amateur boxers. *British Journal of Sports Medicine, 35*, 390–395.

Hall, C. R. (2001). Imagery in sport and exercise. In R. N. Singer, H. A. Hausenblas, & C. M. Janelle (Eds.), *Handbook of sport psychology* (2nd ed., pp. 529–549). New York, NY: Wiley.

Hall, K. D. (2008). What is the required energy deficit per unit weight loss? *International Journal of Obesity (Lond), 32*(3), 573–576.

Hanin, Y. (2000). Individual zones of optimal functioning (IZOF) model: Emotion-performance relationships in sport. In Y. Hanin (Ed.), *Emotions in sport* (pp. 65–89). Champaign, IL: Human Kinetics.

Hanson, T. (2006). Focused baseball: Using sport psychology to improve baseball performance. In J. Dosil (Ed.), *The sport psychologist's handbook: A guide for sport-specific performance enhancement* (pp. 159–182). Chichester, UK: Wiley.

Hanton, S., & Jones, G. (1999). The effects of multi-modal intervention program on performers: Pt. II. Training butterflies to fly in formation. *Sport Psychologist, 13*, 22–41.

Hardy, J., Hall, C. R., & Hardy, L. (2004). A note on athletes' use of self-talk. *Journal of Applied Sport Psychology, 16*, 251–257.

Hardy, J., & Zourbanos, N. (2016). Self-talk in sport: Where are we now? In R. J. Schinke, K. R. McGannon, & B. Smith (Eds.), *Routledge international handbook of sport psychology* (pp. 449–460). Abingdon, Oxon, UK: Routledge.

Harris, D. V., & Harris, B. L. (1984). *The athlete's guide to sport psychology: Mental skills for physical people.* New York, NY: Leisure Press.

Hatzigeorgiadis, A., Zourbanos, N., Latinjak, A. T., & Theodorakis, Y. (2014). Self-talk. In A. G. Papaioannou & D. Hackfort (Eds.), *Routledge companion to sport and exercise psychology: Global perspectives and fundamental concepts* (pp. 372–385). Hove, East Sussex, UK: Routledge.

Hausswirth, C., & Mujika, I. (Ed.). (2013). *Recovery for performance in sport.* Champaign, IL: Human Kinetics.

Hayes, A., & Cribb, P. J. (2008). Effect of whey protein isolate on strength, body composition and muscle hypertrophy during resistance training. *Current Opinion in Clinical Nutrition and Metabolic Care, 11*(1), 40–44.

Heishman, M., & Bunker, L. (1989). Use of mental preparation strategies by international elite female lacrosse players from five countries. *Sport Psychologist, 3*, 138–150.

Helms, E. R., Aragon, A. A., & Fitschen, P. J. (2014). Evidence-based recommendations for natural bodybuilding contest preparation: Nutrition and supplementation. *Journal of the International Society of Sports Nutrition, 11*, 20.

Henschen, K. (2005). Mental practice–Skill oriented. In D. Hackfort, J. Duda, & R. Lidor (Eds.), *Handbook of research in applied sport and exercise psychology: International perspectives* (pp. 19–36). Morgantown, WV: Fitness Information Technology.

Henschen, K., Statler, T., & Lidor, R. (2007). Psychological factors of tactical preparation. In B. Blumenstein, R. Lidor, & G. Tenenbaum (Eds.), *Psychology of sport training* (pp. 104–114). Oxford, UK: Meyer & Meyer Sport.

Henschen, K. H., & Cook, D. (2003). Working with professional basketball players. In R. Lidor & K. P. Henschen (Eds.), *The Psychology of team sports* (pp. 143–160). Morgantown, WV: Fitness Information Technology Publishers.

Hodge, K., & Kentta, G. (2016). Athlete burnout. In R. Schinke, K. McGannon & B. Smith (Eds.), *Routledge international handbook of sport training* (pp. 157–166). New York, NY: Routledge.

Holliday, B., Burton, D., Sun, G., Hammermeister, J., Naylor, S., & Freigang, D. (2008). Building the better mental training mousetrap: Is periodization a more systematic approach to promoting performance excellence? *Journal of Applied Sport Psychology, 20*, 199–219.

Hooper, L., Abdelhamid, A., Moore, H. J., Douthwaite, W., Skeaff, C. M., & Summerbell, C. D. (2012). Effect of reducing total fat intake on body weight: Systematic review and meta-analysis of randomised controlled trials and cohort studies. *BMJ, 345*, e7666.

Houston, M. E. (1999). Gaining weight: The scientific basis of increasing skeletal muscle mass. *Canadian Journal of Applied Physiology, 24*(4), 305–316.

Houtkooper, L., Abbot, J. M., & Nimmo, M. (2007). Nutrition for throwers, jumpers and combined events athletes. *Journal of Sports Sciences, 25*(S1), S39–S47.

Howarth, K. R., Phillips, S. M., MacDonald, M. J., Richards, D., Moreau, N. A., & Gibala, M. J. (2010). Effect of glycogen availability on human skeletal muscle protein turnover during exercise and recovery. *Journal of Applied Physiology, 109*(2), 431–438.

Howell, S. & Kones, R. (2017). "Calories in, calories out" and macronutrient intake: the hope, hype, and science of calories. American Journal of Physiology Endocrinology and Metabolism, 313(5), E608-E612.

Hulthen, L., Bengtsson, B. A., Sunnerhagen, K. S., Hallberg, L., Brimby, G., & Johansson, G. (2001). GH is needed for the maturation of muscle mass and strength in adolescence. *The Journal of Clinical Endocrinology and Metabolism, 86*(10), 4765–4770.

Israetel, M., Hoffmann, J., & Smith, C. (2015). *The scientific principles of strength training.* JTS.

Issurin, V. B. (2013). Training transfer: Scientific background and insights for practical application. *Sports Medicine.* doi:10.1007/s40279-013-0049-6

Ivy, J. (Ed.). (2004). *Nutrient timing: The future of sports nutrition*. Laguna Beach, CA: Basic Health Publications.

Izquierdo, M., Ibanez, J., Gonzales-Badillo, J. J., Ratamess, N. A., Kraemer, W. J., Hakkinen, K., ... Gorostiaga, E. M. (2007). Detraining and tapering effects on hormonal responses and strength performance. *Journal of Strength and Conditioning Research, 21*, 768–775.

Jacobsen, B. H., Sobonya, C., & Ransone, J. (2001). Nutrition practices and knowledge of college varisty athletes: A follow-up. *Journal of Strength and Conditioning Research, 15*(1), 63–68.

Jackson, R. C. (2014). Preperformance routines. In R. C. Eklund & G. Tenenbaum (Eds.), *Encyclopedia of sport and exercise psychology* (pp. 550–553). Thousand Oaks, CA: Sage Publications.

Jackson, S. (2000). Joy, fun and flow state in sport. In Y. Hanin (Ed.), *Emotions in sport* (pp. 135–156). Champaign, IL: Human Kinetics.

Jacobson, E. (1938). *Progressive relaxation.* Chicago, IL: University of Chicago Press.

Jeukendrup, A., & Gleeson, M. (Ed.). (2010). *Sport Nutrition* (2nd ed.). Champaign, IL: Human Kinetics.

Judelson, D. A., Maresh, C. M., Anderson, J. M., Armstrong, L. E., Casa, D. J., Kraemer, W. J., & Volek, J. S. (2007). Hydration and muscular performance: Does fluid balance affect strength, power and high-intensity endurance? *Sports Medicine, 37*(10), 907–921.

Karageorgis, C., & Priest, D. (2012). Music in the exercise domain: A review and synthesis. *International Review of Sport and Exercise Psychology, 5*(1), 44–66. doi:10.1080/1750984x.2011.631026

Keesey, R. E., & Hirvonen, M. D. (1997). Body weight set-points: Determination and adjustment. *Journal of Nutrition, 127*(9), 1875S–1883S.

Kellmann, M., & Beckmann, J. (2003). Research and intervention in sport psychology: New perspectives on an inherent conflict. *International Journal of Sport and Exercise Psychology, 1*, 13–26.

Kerksick, M., Arent, S., Schoenfeld, B. J., Stout, J. R., Campbell, B., Wiborn, C. D....Antonio, J. (2017). International society of sport nutrition position stand: nutrient timing. Journal of the *International Society of Sports Nutrition, 14*, 33.

Kerksick, C, Harvey, T., Stout, J., Campbell, B., Wilborn, C., Kreider, R., ... Antonio, J. (2008). International Society of Sports Nutrition position stand: Nutrient timing. *Journal of International Society of Sports Nutrition, 5*, 17.

Klem, M. L., Wing, R. R., McGuire, M. T., Seagle, H. M., & Hill, J. O. (1998). Psychological symptoms in individuals successful at long-term maintenance of weight loss. *Health Psychology, 17*(4), 336–345.

Krane, V., & Williams J. (2015). Psychological characteristics of peak performance. In J. M. Williams & V. Krane (Eds.), *Applied sport psychology: Personal growth to peak performance* (7th ed., pp. 159–175). New York, NY: McGraw Hill.

Kubukeli, Z. N., Noakes, T. D., & Dennis, S. C. (2002). Training techniques to improve endurance exercise performances. *Sports Medicine, 32*, 489–509.

Kumar, V., Abbas, A. K., & Fausto, N. (2005). *Robbins and Cotran: Pathologic basis of disease (7th Edition).* Philadelphia, PA: Elsevier Saunders.

Laemmle, J., & Martin, B. (2013). Children at play: Learning gender in the early years. *Journal of Youth Adolescent, 42*(2), 305–307.

Lebon, F., Collet, C., & Guillot, A. (2010). Benefits of motor imagery training on muscle strength. *Journal of Strength and Conditioning Research, 24*(6), 1680–1687. doi:10.1519/JSC.0b013e3181d8e936

Leidy, H. J., Clifton, P. M., Astrup, A., Wycherley, T. P., Westerterp-Plantenga, M. S., Luscombe-Marsh, M. S., ... Mattes, R. D. (2015). The role of protein in weight and maintenance. *The American Journal of Clinical Nutrition, 6*(101), 1320s–1329s.

Lidor, R. (2007). Preparatory routines in self-paced events. Do they benefit the skilled athletes? Can they help the beginners? In G. Tenenbaum & R. Eklund (Eds.), *Handbook of sport psychology* (3rd ed., pp. 445–468). Hoboken, NJ: Wiley.

Lidor, R., Blumenstein, B., & Tenenbaum, G. (2007). Psychological aspects of training in European basketball: Conceptualization, periodization, and planning. *The Sport Psychologist, 21*, 353–367.

Loehr, J. E. (1986). *Mental toughness training for sports: Achieving athletic excellence.* New York, NY: The Stephen Greene Press.

Logan, J. G., & Barksdlae, D. J. (2008). Allostasis and allostatic load: Expanding the discourse on stress and cardiovascular disease. *Journal of Clinical Nursing, 17*(7b), 201–208.

Loucks, A. B. (2004). Energy balance and body composition in sports and exercise. *Journal of Sports Sciences, 22*(1), 1–14.

Loucks, A. B., Kiens, B., & Wright, H. H. (2011). Energy availability in athletes. *Journal of Sports Science, 29*, S7–S15.

Manore, M. M. (2012). Dietary supplements for improving body composition and reducing body weight: Where is the evidence? *International Journal of Sport Nutrition and Exercise Metabolism, 22*(2), 139–154.

Marchant, D. C., Grieg, M., Bullough, J., & Hitchen, D. (2011). Instructions to adopt an external focus enhance muscular endurance. *Research Quarterly for Exercise and Sport Science, 82*(3), 466–473.

Margaritis, I, Palazetti, S, Rousseau, A.-S, Richard, M.-J., & Favier, A. (2003). Antioxidant supplementation and tapering exercises improve exercise-induced antioxidant response. *Journal of American College of Nutrition, 22*(2), 147–156.

Martin, K. A., & Hall, C. R. (1995). Using mental imagery to enhance intrinsic motivation. *Journal of Sport and Exercise Psychology, 17*, 54–69.

Maughan, R. J., & Shirreffs, S. M. (2010). Development of hydration strategies to optimize performance for athletes in high-intensity sports and in sports with repeated intense efforts. *Scandinavian Journal of Medicine & Science in Sports, 20*(s2), 59–69.

McNair, D., Lorr, M., & Droppleman, L. (1971). *Profile of mood states manual.* San Diego, CA: Educational and Testing Service.

McNeely, E., & Sadler, D. (2007). Tapering for endurance athletes. *Strength and Conditioning Journal, 29*, 18–24.

Meyers, A., Whelan, J., & Murphy, S. (1996). Cognitive behavior strategies in athletic performance enhancement. In M. Mersen, R. Miller, & A. Belack (Eds.), *Progress in behavior modification* (pp. 137–164). Pacific Grove, CA: Brooks/Cole.

Mondazzi, L., & Arcelli, E. (2009). Glycemic index in sport nutrition. *Journal of the American College of Nutrition, 28*, 455S–463S.

Moran, A. (2003). Improving concentration skills in team-sport performance: Focusing techniques for soccer players. In R. Lidor & K. Henschen (Eds.), *The psychology of team sports* (pp. 161–190). Morgantown, WV: Fitness Information Technology.

Moran, A. (2010). Concentration/attention. In S. J. Hanrahan & M. B. Andersen (Eds.), *Routledge handbook of applied sport psychology: A comprehensive guide for students and practitioners* (pp. 500–509). Abingdon, Oxon, UK: Routledge.

Moran, A. P. (1996). *The psychology of concentration in sport performers: A cognitive analysis.* East Sussex, UK: Psychology Press.

Morris, T. (2010). Imagery. In S. J. Hanrahan & M. B. Andersen (Eds.), *Routledge handbook of applied sport psychology: A comprehensive guide for students and practitioners* (pp. 481–490). Abingdon, Oxon, UK: Routledge.

Morris, T., Spittle, M., & Watt, A. P. (2005). *Imagery in sport.* Champaign, IL: Human Kinetics.

Morton, R. W., McGlory, C., & Phillips, S. M. (2015). Nutritional interventions to augment resistance training-induced skeletal muscle hypertrophy. *Frontiers in Physiology, 6*, 245.

Mujika, I. (2007). Challenges of team sport research. *International Journal of Sports Physiology and Performance, 2*, 221–222.

Mujika, I. (2009). *Tapering and peaking for optimal performance.* Champaign, IL: Human Kinetics.

Mujika, I., Chatard, J. C., Padilla, S., Guezennec, C. Y., & Geyssant, A. (1996). Hormonal responses to training and its tapering off in competitive swimmers: Relationship with performance. *European Journal of Applied Physiology and Occupational Physiology, 74*(4), 361–366.

Mujika, I., Goya, A., Padilla, S., Grijalba, A., Gorostiaga, E., & Ibanez, J. (2000). Physiological responses to a 6-d taper in middle-distance runners: Influence of training intensity and volume. *Medicine and Science in Sports and Exercise, 32*(2), 511–517.

Mujika, I., Goya, A., Ruiz, E., Grijalba, A. Santisteban, J., & Padilla, S. (2002). Physiological and performance responses to a 6-day taper in middle-distance runners: Influence of training frequency. *International Journal of Sports Medicine, 23*(5), 367–373. doi:10.1055/s-2002-33146

Mujika, I., & Padilla, S. (2003). Scientific basis for precompetition tapering strategies. *Medical Science of Sport Exercise, 35*(7), 1182–1187.

Mujika, I., Padilla, S., Pyne, D., & Busso, T. (2004). Physiological changes associated with the pre-event taper in athletes. *International Journal of Sports Medicine, 34*(13), 891–927.

Neary, J. P., Martin, T. P., & Quinney, H. A. (2003). Effects of taper on endurance cycling capacity and single muscle fiber properties. *Medicine and Science in Sports and Exercise, 35*, 1875–1881.

Nederhof, E., Lemmink, K. A. P. M., Zwerver, H. J., & Mulder, T. (2007). The effect of high load training on psychomotor speed. *International Journal of Sports Medicine, 28*, 595–601.

Oliver, J. (2010). Ethical practice in sport psychology: Challenges in the real world. In S. J. Hanrahan & M. B. Andersen (Eds.), *Routledge handbook of applied sport psychology: A comprehensive guide for students and practitioners* (pp. 60–68). Abingdon, Oxon, UK: Routledge.

O'Reilly, J., Wong, S. H., & Chen, Y. (2010). Glycaemic index, glycaemic load and exercise performance. *Sports Medicine, 40*(1), 27–39.

Orlick, T. (1990). *In pursuit of excellence.* Champaign, IL: Leisure Press.

Orlick, T., & Partington, J. (1988). Mental links to excellence. *The Sport Psychologist, 2*, 105–130.

Patrick, T. D., & Hrycaiko, D. W. (1998). Effects of a mental training package on an endurance performance. *The Sport Psychologist, 12*, 283–299.

Phillips, S. M. (2012). Dietary protein requirements and adaptive advantages in athletes. *British Journal of Nutrition, 108*(2), S158–S167.

Phillips, S. M., Moore, D. R., & Tang, J. E. (2007). A critical examination of dietary protein requirements, benefits, and excesses in athletes. *International Journal of Sport Nutrition and Exercise Metabolism, 17*, S58–S76.

Phillips, S. M., & Van Loon, L. J. (2011). Dietary protein for athletes: From requirements to optimum adaptation. *Journal of Sports Science, 29*, S29–S38.

Prasad, D. C., & Das, B. C. (2009). Physical inactivity: A cardiovascular risk factor. *Indian. Journal of Medical Science, 63*(1), 33–42.

Raglin, J. S., Koceja, D. M., & Stager, J. M. (1996). Mood, neuromuscular function and performance during training in female swimmers. *Medicine and Science in Sports and Exercise, 28*, 372–377.

Ravizza, K. (1977). Peak experiences in sport. *Journal of Humanistic Psychology, 17*, 35–40.

Reiser, M., Busch, D., & Munzert, G. (2011). Strength gains by motor imagery with different ratios of physical to mental practice. *Frontiers in Psychology, 2*(194). doi:10.3389/fpsyg.2011.00194

Rietjens, G. J. W. M., Keizer, H. A., Kuipers, H., & Saris, W. H. (2001). A reduction in training volume and intensity for 21 days does not impair performance in cyclists. *British Journal of Sports Medicine, 35*, 431–434.

Rogol, A. D., Roemich, J. N., & Clark, P. A. (2002). Growth at puberty. *Journal of Adolescent Health, 31*(6), S192–S200.

Sahlin, K. (2014). Muscle energetics during explosive activities and potential effects of nutrition and training. *Sports Medicine, 44*, S167–S173.

Samuels, C., James, L., Lawson, D., & Meeuwissee, W. (2016). The athlete sleep screening questionnaire: A new tool for assessing and managing sleep in elite athletes. *British Journal of Sports Medicine, 50*(7), 418. doi:10.1136/bjsports-2014-094332

Schaafsma, G. (2000). The protein digestibility-corrected amino acid score. *Journal of Nutrition, 130*(7), 1865S–1867S.

Schultz, J. (1932). *Das autogene Training* (Autogenic training). Stuttgart, Germany: Thieme.

Selye, H. (1984). *The stress of life.* New York, NY: McGraw-Hill.

Sheard, M., & Golby, J. (2006). Effect of a psychological skills training program on swimming performance and positive psychological development. *International Journal of Sport and Exercise Psychology, 4*(2), 149–169.

Shepley, B, MacDoughall, J. D., Cipriano, N, Sutton, J. R., Tarnopolsky, M. A., & Coates, G. (1992). Physiological effects of tapering in highly trained athletes. *Journal of Applied Physiology, 72*, 706–711.

Shi, X., & Gisolfi, C. V. (1998). Fluid and carbohydrate replacement during intermittent exercise. *Sports Medicine, 25*(3):157–172.

Si, G., Statler, T., & Samulski, D. (2014). Preparing athletes for major competition. In A. G. Papaioannou & D. Hackfort (Eds.), *Routledge companion to sport and exercise psychology: Global perspectives and fundamental concepts* (pp. 495–510). Hove, East Sussex, UK: Routledge.

Singer, R. N., & Anshel, M. (2006). An overview of interventions in sport. In J. Dosil (Ed.), *The sport psychologist handbook–A guide for sport-specific performance enhancement* (pp. 63–88). West Sussex, UK: John Wiley & Sons.

Slater, G., & Phillips, S. M. (2011). Nutrition guidelines for strength sports: Sprinting, weightlifting, throwing events, and bodybuilding. *Journal of Sports Science, 29*(1), S67–S77.

Smith, D., Collins, D., & Holmes, P. (2003). Impact and mechanism of mental practice effects on strength. *International Journal of Sport Psychology, 1*, 293–306.

Smith, R. E. (1989). Applied sport psychology in an age of accountability. *Journal of Applied Sport Psychology, 1*, 166–180.

Spriet, L. L. (2014). Nutrition for training and performance. *Sports Medicine, 44*, S115–S116.

Stambulova, N., & Wylleman, P. (2014). Athletes' career development and transitions. In A. Papaioannou & D. Hackfort (Eds.), *Routledge companion to sport and exercise psychology: Global perspective and fundamental concepts* (pp. 605–620). Hove, East Sussex: Routledge.

Statler, T., & Henschen, K. (2009). A sport psychology service delivery model for developing and current track and field athletes and coaches. In T. Hung, R. Lidor, & D. Hackfort (Eds.), *Psychology of sport excellence* (pp. 25–31). Morgantown, WV: Fitness Information Technology.

Sterling, P. (2004). *Principles of allostasis: Optimal design, predictive regulation, pathophysiology and rational therapeutics.* Cambridge: Cambridge University Press.

Stiegler, P., & Cunliffe, A. (2006). The role of diet and exercise for the maintenance of fat-free mass and resting metabolic rate during weight loss. *Sports Medicine, 36*(3), 239–262.

Suinn, R. (1993). Imagery. In R. Singer, M. Murphey, & L. Tennant (Eds.), *Handbook of research on sport psychology* (pp. 492–510). New York, NY: Macmillan.

Sumithran, P., & Proietto, J. (2013). The defence of body weight: A physiological basis for weight regain after weight loss. *Clinical Science (Lond), 124*(4), 231–241.

Taylor, J. L., & Gandevia, S. C. (2008). Fatigue mechanisms determining exercise performance: A comparison of central aspects of fatigue in submaximal and maximal voluntary contractions. *Journal of Applied Physiology, 104*, 542–550.

Thelwell, R., & Greenlees, I. A. (2001). The effect of a mental skills training package on gymnasium triathlon performance. *The Sport Psychologist, 15*, 127–141.

Thelwell, R. C., & Greenlees, I. A. (2003). Developing competitive endurance performance using mental skills training. *The Sport Psychologist, 17*, 318–337.

Thomas, L., & Busso, T. (2005). A theoretical study of taper characteristics to optimise performance. *Medicine and Science in Sports and Exercise, 37*, 1615–1621.

Tod, D. A., Iredale, K. F., McGuigan, M. R., Strange, D. E., & Gill, N. (2005). "Psyching-Up" enhances force production during the bench press exercise. *Journal of Strength and Conditioning Research, 19*(3), 599–603.

Tortora, G. J., & Derrickson, B. (2012). *Principles of anatomy and physiology (13th Edition).* Hoboken, NJ: John Wiley and Sons.

Trappe, S., Costill, D., & Thomas, R. (2001). Effect of swim taper on whole muscle and single fiber contractile properties. *Medicine and Science in Sports and Exercise, 33*, 48–56.

Trinity, J. D., Pahnke, M. D., Reese, E. C., & Coyle, E. F. (2006). Maximum mechanical power during taper in elite swimmers. *Medicine and Science in Sports and Exercise, 38*, 1643–1649.

Tuomilehto, H., Vuorinen, V. P., Penttila, E., Kivimaki, M., Vuorenmaa, M., Venojarvi, M., ... Pihlajamaki, J. (2017). Sleep of professional athletes: Underexploited potential to improve health and performance. *Journal of Sport Sciences, 35*(7), 704-710.

Vealey, R. (1988). Future directions in psychological skills training. *Sport Psychologist, 2*, 318–336.

Vealey, R. (1994). Current status and prominent issues in sport psychology interventions. *Medicine and Science in Sport and Exercise, 26*, 318–336.

Vealey, R. (2005). *Coaching for the inner edge.* Morgantown, WV: Fitness Information Technology.

Vealey, R. (2007). Mental skills training in sport. In G. Tenenbaum & R. C. Eklund (Eds.), *Handbook of sport psychology* (3rd ed., pp. 287–309). New York, NY: Wiley.

Vealey, R., & Forlenza, R. (2015). Understanding and using imagery in sport. In J. M. Williams & V. Krane (Eds.), *Applied sport psychology: Personal growth to peak performance* (7th ed., pp. 240–273). New York, NY: McGraw Hill.

Vealey, R., & Greenleaf, C. (2006). Seeing is believing: Understanding and using imagery in sport. In J. Williams (Ed.), *Applied sport psychology: Personal growth to peak performance* (5th ed., pp. 306–348). Boston, MA: McGraw-Hill.

Vealey, R. S., & Vernau, D. (2010). Confidence. In S. J. Hanrahan & M. B. Andersen (Eds.), *Routledge handbook of applied sport psychology: A comprehensive guide for students and practitioners* (pp. 518–527). Abingdon, Oxon, UK: Routledge.

Vernacchia, R. (2003). Working with individual team sports: The psychology of track and field. In R. Lidor & K. Henschen (Eds.), *The psychology of team sports* (pp. 235–265). Morgantown, WV: Fitness Information Technology.

Vollaard, N. B., & Shearman, J. P. (2006). Exercise-induced oxidative stress in overload training and tapering. *Medicine and Science in Sports and Exercise, 38*, 1335–1341.

Weinberg, R. (2010). Activation/arousal control. In S. Hanrahan & M. Andersen (Eds.), *Routledge handbook of applied sport psychology* (pp. 471–480). Abingdon, UK: Routledge.

Weinberg, R., & Butt, J. (2014). Goal-setting in sport performance. In A. G. Papaioannou & D. Hackfort (Eds.), *Routledge companion to sport and exercise psychology: Global perspectives and fundamental concepts* (pp. 343–355). Hove, East Sussex, UK: Routledge.

Weinberg, R., & Comar, W. (1994). The effectiveness of psychological interventions in competitive sport. *Sport Medicine, 18*, 406–418.

Weinberg, R. S. (1988). *The mental advantage: Developing your psychological skills in tennis*. Champaign, IL: Leisure Press.

Weinberg, R. S., & Gould, D. (2015). *Foundations of sport and exercise psychology* (6th ed.). Champaign, IL: Human Kinetics.

Weinberg, R. S., & Williams, J. M. (2015). Integrating and implementing a psychological skills training program. In J. M. Williams & V. Krane (Eds.), *Applied sport psychology: Personal growth to peak performance* (7th ed., pp. 329–358). New York, NY: McGraw Hill.

Weinsier, R. L., Nagy, T. R., Hunter, G. R., Darnell, B. E., Hensrud, D. D., & Weiss, H. L. (2000). Do adaptive changes in metabolic rate favor weight regain in weight-reduced individuals? An examination of the set-point theory. *American Journal of Clinical Nutrition, 72*(5), 1088–1094.

Westerterp-Plantenga, M. S., Nieuwenhuizen, A., Tomé, D., Soenen, S., & Westerterp, K. R. (2009). Dietary protein, weight loss, and weight maintenance. *Annual Review of Nutrition, 29*, 21–41.

Wild, C. Y., Steele, J. R., & Munro, B. J. (2013). Musculoskeletal and estrogen changes during the adolescence growth spurt in girls. *Medicine and Science in Sports and Exercise, 45*(1), 97–109.

Wrisberg, C. A., & Pein, R. L. (1990). Past running experience as a mediator of the attentional focus of male and female recreational runners. *Perceptual and Motor Skills, 70*, 427–432.

Wylleman, P., Rosier, N., & De Knop, P. (2016). Career transition. In R. Schinke, K. McGannon & B. Smith (Eds.), *Routledge international handbook of sport psychology* (pp. 111-118). New York, NY: Routledge.

Yao, W. X., Ranganathan, V. K., Allexandre, D., Siemionow, V., & Yue, G. H. (2013). Kinesthetic imagery training of forceful muscle contractions increases brain signal and muscle strength. *Frontiers in Human Neuroscience, 7*(561). doi:10.3389/fnhum.2013.00561

Zaichkowsky, L. D., & Fuchs, C. Z. (1988). Biofeedback application in exercise and athletic performance. In K. B. Pandolf (Ed.), *Exercise and sports science reviews* (pp. 381–421). New York, NY: Macmillan.

Zarkardas, P. C., Carter, J. B., & Banister, E. W. (1995). Modeling the effect of taper on performance, maximal oxygen uptake, and the anaerobic threshold in endurance triathletes. *Advances in Experimental Medicine and Biology, 393*, 179–186.

Zatsiorsky, V. (1995). *Science and practice of strength training*. Champaign, IL: Human Kinetics.

Ziegler, E. F. (1987). Rationale and suggested dimensions for an ethics code for sport psychologists. *The Sport Psychologist, 1*, 138–150.

Ziv, G., & Lidor, R. (2009). Physical attribute, physiological characteristics, on-court performances and nutritional strategies of female and male basketball players. *Sports Medicine, 39*(7), 547–568.

Zoorob, R., Parrish, M. E., O'Hara, H., & Kalliny, M. (2013). Sports nutrition needs: Before, during, and after exercise. *Primary Care, 40*(2), 475–486.

ABOUT THE AUTHORS

TUDOR BOMPA

The Romanian-born Tudor Olimpius Bompa, PhD, has been captivated by training theories from the early 1960s, such as planning-periodization and periodization of dominant biomotor abilities. Some of his theories have been applied in sports training, military training, bodybuilding, and fitness.

Professor Bompa has published 16 books—12 on training theories, 1 on social anthropology, and 3 in political science—in the United States, Canada, Germany, Poland, and Romania. Some of his books have been translated into 19 languages and are used as textbooks in over 180 countries. His book *Periodization: Theory and Methodology of Training* is one of the most popular textbooks in universities worldwide. He has also made presentations in 46 countries around the world.

Tudor Bompa has received 23 distinctions from 21 countries, including professor emeritus from York University, Toronto, Ontario, Canada, Lifelong Achievement Award from NSCA (Las Vegas, 2014), and Doctor Honoris Causa from Polytechnic University, Timisoara, Romania (2017).

BORIS BLUMENSTEIN (ED.)

Prof. Boris Blumenstein is the director of the Department of Behavioral Sciences at the Ribstein Center for Sport Medicine Sciences and Research at the Wingate National Institute of Sport; he is also the associate professor of the MA program in sport and exercise psychology at the College of Management, Academic Studies, Rishon Lezion, and at Givat Washington Teachers College (MA program). He received his PhD in sport psychology in 1980 from the All Union Institute for Research in Sport, Department of Sport Psychology, Moscow, Russia (former USSR).

Prof. Blumenstein's extensive experience in sport psychology spans some 30 years, culminating in applied work at the elite level. He has been a sport psychology consultant for the Soviet national and Olympic teams and since 1990 to the Israeli national and Olympic teams (including delegations to Atlanta, 1996; Sydney, 2000; Athens, 2004; and Beijing, 2008). He is author and coauthor of seven books, 29 book chapters, and over 60 refereed journal articles, mainly in area of sport and exercise psychology. He has also made more than 80 scientific presentations at international and national conferences and workshops.

Prof. Blumenstein's current research interest includes mental skills training for performance, stress-performance relationship, and effectiveness of different mental interventions for athletic competitions readiness. In addition, he is past president of the Israeli Society for Sport Psychology.

JAMES HOFFMANN

Dr. James Hoffmann holds a PhD in sport physiology from East Tennessee State University. He earned his PhD under Dr. Mike Stone, where he focused on the application of sled pushing to sport performance enhancement in Rugby players. James is the former program director of the exercise and sport sciences program at Temple University and is currently a sports performance consultant for Renaissance Periodization. James has taught numerous courses on strength and conditioning, nutrition, and exercise physiology.

James has coauthored numerous books, including *The Renaissance Diet, The Scientific Principles of Strength Training, How Much Should I Train: An Introduction to the Volume Landmarks*, and *Recovering from Training: How to Manage Fatigue to Maximize Performance*. James and his colleagues have also performed dozens of seminars around the world in the theory and practices of training, nutrition, and recovery strategies.

James has coached thousands of athletes and fitness enthusiasts in the areas of fitness, nutrition, and recovery. He has served as a Division 1 strength and conditioning coach and has also coached men and women's Rugby at the collegiate level. As a lifelong athlete himself, James has achieved high ranks in competitive Rugby, American football, and wrestling, and he is currently pursuing Thai boxing.

SCOTT HOWELL

Dr. Scott Howell is the international director of Tudor Bompa Institute (TBI). He holds a MD and PhD in exercise physiology. Dr. Howell has a keen interest in research methods, statistics, epidemiology, physiology, and sports science. He is a renowned specialist of periodization and its application in sports training. As the international director of TBI, Dr. Howell has promoted TBI internationally, with national offices in 36 countries.

Teaching, particularly in the areas of sports science and periodization, is one of Dr. Howell's passions. He is a frequent guest lecturer at several institutions in the United States on subjects such as statistics; adverse effects of androgens related to cardiovascular, liver, and kidney disease; toxicology of performance-enhancing drugs; strength and fitness; epidemiology; nutrition; and the principles of periodization.

Dr. Scott Howell has published scientific papers on a variety of topics in several academic journals, including the *American Journal of Physiology*, *Journal of Obesity and Diabetes*, *Yale Journal of Biology and Medicine*, and *Karger (Cardiology)*. He has received several certifications from medical and sports-related organizations. Dr. Howell is a member of several medical, research, and sport training organizations.

IRIS ORBACH (ED.)

Dr. Iris Orbach is a researcher and a sport psychology consultant in the Department of Behavioral Sciences at the Ribstein Center for Sport Medicine Sciences and Research, Wingate National Institute of Sport; head of the MA program in sport and exercise psychology, the College of Management, Academic Studies, Rishon Lezion; and head of Health Promotion Program at the Nat Holman School for Coaches and Instructors, Wingate National Institute of Sport.

Dr. Orbach received her PhD in sport psychology in 1999 from the University of Florida, Department of Sport and Exercise Sciences, in Gainesville, Florida, USA. She worked as an assistant professor in the Department of Sport, Fitness, and Leisure Studies at Salem State University, Salem, Massachusetts, USA. In addition to teaching, Dr. Orbach has published numerous articles and book chapters and has given presentations at national and international conferences on topics related to sport psychology. She is the coauthor of the books *Mental Practice in Sport: Twenty Case Studies* and *Psychological Skills Training* published by Nova Science Publishers, 2012.

Dr. Orbach's current research interests include stress-performance relationship, children and motivation in sport, and the effectiveness of various mental training practices. Dr. Orbach uses her psychology skills as a consultant for athletes at all skill levels. In her free time, Dr. Orbach enjoys running, bicycling, swimming, weight lifting, and all kinds of fitness activities.